APPROACHES TO QUANTITATIVE RESEARCH

A GUIDE FOR DISSERTATION STUDENTS

Edited by

Helen Xiaohong Chen

Published by
OAK TREE PRESS
19 Rutland Street, Cork, Ireland
www.oaktreepress.com

A catalogue record of this book is
available from the British Library.

ISBN: 978-1-78119-058-6 (Wirebound)
ISBN: 978-1-78119-059-3 (Paperback)
ISBN: 978-1-78119-060-9 (ePub)
ISBN: 978-1-78119-061-6 (Kindle)

CONTENTS

FIGURES

TABLES

CONTRIBUTORS

Helen Chen, BA, MBA, PhD, has lectured in the University of Warwick (UK), City University of Hong Kong, University of Kent (UK), and Carleton University (Canada), before joining the Dublin Institute of Technology (Ireland). She lectures in international marketing and marketing modules. Her research interests include consumer behaviour, international marketing and foreign market entry modes. She has presented at conferences such as the Academy of International Business and the Irish Academy of Management. She has published in *Journal of Organizational Computing and Electronic Commerce, Journal of Marketing Science,* and *Irish Business Journal.*

Kieran Flanagan, BSc, MSc, DipCMS, MICS, MSc, is a lecturer in Information Systems and Management Science in the Business Faculty at the Dublin Institute of Technology. He has qualifications in Mathematics, Mathematical Physics, Computer Modelling and Simulation and Applied Computing. Kieran is also a member of the Irish Computer Society. He has presented a paper on soft computing approaches to forecasting at the conference on applied statistics in Ireland (CASI). His research interests lie in the area of forecasting, with particular emphasis on soft computing and statistical techniques.

Daire Hooper, BSc, PhD, is a lecturer at the Dublin Institute of Technology. Daire teaches research methods, marketing research and services marketing theory courses. Her research interests include services marketing, servicescapes, consumption emotions and behavioural responses to consumer experiences. She has presented at numerous international conferences, including the *American Marketing Association SERVSIG Services Research Conference,*

QUIS11 and the *European Conference on Research Methodology for Business and Management Studies*. Her work has been published in the *Electronic Journal of Business Research Methods*.

Agnes Maciocha, MSc, PhD, is a lecturer in Business at the Institute of Art, Design and Technology, Dun Laoghaire. Her research interests deal with application of quantitative methods and data-mining techniques for decision-making in business, with particular emphasis on the role of intangible assets and intellectual capital measurement in evaluating economic performance of companies. She has presented at numerous conferences, including the International Conference on Applied Statistics, the Irish Academy of Management, and the European Conference on Intellectual Capital. Her work has been published in *Lecture Notes on Computer Science, Electronic Journal of Knowledge Management,* and *Irish Journal of Management*.

Siobhan Mc Carthy, MLitt, BA(Mod.), is an economist who has taught in the Economics Department of Trinity College Dublin and NUI Maynooth and currently teaches in the Dublin Institute of Technology. She teaches Economy of Ireland, European Economics, and Quantitative Methods. Her research interests include environmental and resource economics, in particular emissions trading regimes, establishing econometrical models for the voluntary sector and third level education participation and work/life balance issues. She has published in the *Land Use Policy Journal, The Economic and Social Review* and *Irish Banking Review*. Siobhan and her co-authors have presented conference papers in Ireland, UK and France. Siobhan was awarded a Walsh Fellowship Award from Teagasc to carry out her MLitt research. Siobhan developed research topics for masters' studentships and wrote research proposals to obtain both internal Dublin Institute of Technology funding and external funding for these projects.

Dónal O'Brien BA, MSc, is a lecturer and researcher in the Dublin Institute of Technology. He is currently pursuing a PhD in the area of strategy development in multinational subsidiaries. His papers have been published by the Academy of International Business and, in 2011, he received the award for the best postgraduate paper from the Irish Academy of Management. Prior educational qualifications include a degree in Management and Marketing and a first class master's in Strategic Management. Before being awarded the position in the Dublin Institute of Technology, Dónal gained commercial experience in a number of sectors including banking, accounting, retail and sports and leisure. Dónal has lectured on both full-time and executive courses in the Dublin Institute of Technology.

Eddie Rohan, BSocSc, MSc, PhD, has been interested in research methods and statistics since his undergraduate days. He lectures in the Dublin Institute of Technology and has been a Chief Examiner with the Marketing Institute since 1981. His PhD was in the area of family decision-making, with a special focus on financial decisions within dual career couples. He has PhD students in areas as diverse as inertia in financial services, location of retail outlets and social norms regarding alcohol consumption. He co-authored a chapter on Market Segmentation for direct and database marketers. Currently, he is director of the MSc in Marketing programme in the Dublin Institute of Technology and teaches Marketing Data Analysis and Statistical Methods. Having lectured in marketing research methods at undergraduate, postgraduate and executive levels for many years, Eddie espouses the use of both quantitative as well as qualitative approaches.

Donncha Ryan, MSc, MA, MSc, has a background in Mathematical Physics, Statistics and Financial and Industrial Mathematics. He taught previously in the Institute of Technology Sligo and served as an Examiner for a Management Science module in the Dublin City University Oscail programme. He previously worked as a statistician in the Central Statistics Office and currently works in the Dublin Institute of Technology, lecturing primarily in Marketing Analysis, Management Science and Data Analysis on the Marketing degree programme. He co-authored a chapter on Data Analysis for direct and database marketers. He is particularly interested in the promotion of analytical marketing approaches.

Pamela Sharkey Scott, PhD, FCCA, MBS, was appointed Research Fellow at the Faculty of Business in 2007, following several other academic positions within the Dublin Institute of Technology. She graduated with a PhD in Strategic Planning and Entrepreneurship from University College Dublin in 2006, where she had also obtained a Master's degree. Pamela studied for a professional accounting qualification (FCCA) at the Dublin Business School, following completion of a scholarship from the Governor and Company of the Bank of Ireland to obtain her primary degree. Prior to entering academia, she spent more than 10 years as a Senior Corporate Banker in Bank of Ireland, and has since been involved in several consulting projects. She teaches Strategic Management and International Business at both primary and post-graduate level. Pamela has won several hundred thousand euro in research grants, and has been ranked within the top two positions for the Technology Strand 1 Funding Programme for the last three years. She has published in international journals, including *Strategy and Leadership*, in addition to several book chapters, and is currently co-editing a book on *Irish Subsidiary Management Practices*. Pamela has presented at conferences worldwide and was an invited guest at the 10th Annual IB Research Forum. Her current research interests include strategy development processes, entrepreneurship and developing dynamic capabilities, in an international business context.

ACKNOWLEDGEMENTS

First and foremost, we would like to thank Brian O'Kane of Oak Tree Press for his enormous patience and generous support for the project.

We would like to gratefully acknowledge the support from the College of Business, Dublin Institute of Technology. We wish to express our sincere thanks to Paul O'Sullivan, Director of the College of Business; Kate Uí Ghallachóir, Head of the School of Marketing; Joseph Coughlan, Head of the School of Accounting and Finance; and Paul O'Reilly, Head of Research, for their invaluable support and guidance.

We are grateful to our families and colleagues for their understanding and endless love throughout the project.

The editors and contributors
Dublin
March 2012

CHAPTER 1
INTRODUCTION

Helen Xiaohong Chen

APPROACHES TO QUANTITATIVE RESEARCH is designed for both undergraduate and postgraduate business students who are planning to undertake a research project or dissertation. Due to a lack of formal research training and experience, students can find completing research projects a daunting task. This, coupled with a fear of statistics, can culminate in quite an overwhelming experience for many students. Therefore, we aimed to produce a text that takes a practical approach to quantitative research techniques by providing step-by-step guides to their application and interpretation. By using easy-to-understand language, while at the same time not losing the statistical underpinnings, this book demonstrates how to use the appropriate quantitative methods to answer different types of research questions, and how to analyse data by using SPSS.[1]

ABOUT EACH CHAPTER

A key feature of the book is that the chapters demonstrating techniques for testing hypotheses are all written on the same theme, namely perceived customer satisfaction towards a fictitious clothes retailer, CTRL. The data set[2] used was created by one of the authors, and is used to demonstrate different techniques across the chapters.

---

[1] Statistical Package for the Social Sciences (SPSS) is a commonly-used software package for quantitative investigations among social scientists.

[2] All the variables in the data set are measured by one single item (question). If variables are measured by multiple items, use techniques such as factor analysis to reduce data after preliminary data analysis (see **Chapter 8**).

In the dataset, the dependent variables are customer satisfaction towards CTRL, coded as '**satisfaction**', and the likelihood to shop in CTRL in the next week, coded as '**likelihood**'. The independent variables are waiting time at the check-out, coded as '**waitingtime**'; the background music played in CTRL, coded as '**backgdmusic**'; and the merchandise quality, coded as '**quality**'. Note that, in the literature, more variables should be included to understand customer satisfaction towards a retailer. But in this book, only three independent variables are included for ease of discussion.

Each chapter contains two sections. The first is designed to help students strengthen their knowledge of quantitative techniques and thus it concentrates on the technique itself covering:

- ♦ What the technique is;
- ♦ When it should be used; and
- ♦ The data requirements for using it.

The second section demonstrates how the technique that is the topic of the chapter would be applied in an academic research project, in order to answer a specific research question.

THE STRUCTURE OF THE BOOK

APPROACHES TO QUANTITATIVE RESEARCH does not consider why a quantitative approach should be adopted over a qualitative approach. For such epistemological and methodological discussions, refer to Easterby-Smith, Thorn & Lowe (2002). The starting point of this book is what a researcher needs to do once a quantitative approach has been chosen, given their particular research questions.

Chapter 2, by Eddie Rohan, presents the basics about data and data coding by using SPSS. Different types of data, along with their scales and measurements, may require different types of analytical techniques.

Helen Xiaohong Chen wrote **Chapter 3** to demonstrate how preliminary analyses can be conducted for the hypotheses testing chapters: **Chapters 4, 5, 6** and **7**. The analyses include:

- Descriptives to examine outliers and normal distribution;
- Correlations between variables to examine the collinearity between them; and
- Multivariate normality.

Chapter 4, written by Agnes Maciocha, looks at two main types of t-tests – paired sample t-tests and independent sample t-tests – based on the different characteristics of the data.

Paired sample t-tests are used when a researcher wants to compare the same sample but at two different times. A typical research question associated with perceived customer satisfaction towards a clothes retailer would be:

Q1. What are the differences in perceived customer satisfaction towards CTRL before and after the new management was appointed?

The same participants need to demonstrate their perceived satisfaction towards the retailer at two times: before and after the new management's appointment. Thus, the questionnaire needs to be administered at two different times to the same respondents. Due to lack of data, this particular technique will not be discussed in this book.

Independent sample t-tests are used to compare two groups in the same data set, where the variable being investigated is measured using interval/ratio data. A typical research question associated with customer satisfaction would be:

Q2. What is the difference in satisfaction perceived by female and male customers towards CTRL?

Chapter 4 also examines one-way Analysis of Variance (ANOVA), which is used to compare the differences between two or more groups, when the variable the researcher wants to compare is continuous. Typical research questions associated with perceived customer satisfaction towards a clothes retailer would be:

Q3a. Is there a difference in customer satisfaction towards CTRL as perceived by female and male customers?

Q3b. Is there any difference in customer satisfaction towards CTRL as perceived by customers on lower, medium or higher levels of income?

One-way ANOVA only identifies whether there is a difference between groups; unlike t-tests it does not identify the nature of the difference.

Like t-tests, there are two types of one-way ANOVA: repeated measures ANOVA and between-groups ANOVA.

Two-way ANOVA allows a researcher to include two independent variables for one dependent variable. The advantage of using two-way ANOVA is to understand the interaction effect of two independent variables. A typical research question would be:

Q4. Is there a difference in perceived customer satisfaction towards CTRL between female consumers with lower, medium and higher levels of income and male consumers with lower, medium and higher levels of income?

Similarly, there are two types of two-way ANOVA: repeated measures two-way ANOVA and between-groups two-way ANOVA.

Chapter 5, written by Kieran Flanagan, examines MANOVA, which allows a researcher to compare groups based on a number of different variables that are in a relationship of cause (independent variables) and effect (dependent variables). MANOVA is suitable where more than one dependent variable is under investigation. A typical research question associated with customer satisfaction towards a clothes retailer would be:

Q5. What are the differences in perceived customer satisfaction towards CTRL and the likelihood to shop in CTRL in the next week between female and male customers?

Like ANOVA, MANOVA can be used in a one-way, two-way or higher factorial design fashions. The chapter has an emphasis on factorial MANOVA.

In **Chapter 6**, Dónal O'Brien and Pamela Sharkey Scott look at correlations and regressions. To investigate the relationship between two continuous variables, Pearson correlation and Spearman correlation are the most common techniques used. Both indicate the type of relationship (positively or negatively correlated) and the strength of the relationship (the significance level). A typical research question associated with customer satisfaction towards a clothes retailer would be:

Q6. Does the background music played in the shop affect the perceived customer satisfaction towards CTRL?

Correlation and simple regression often are too simplistic to answer a research question in a research project or dissertation. For example, to better understand customer satisfaction towards a clothes retailer, other variables such as the perceived quality of the clothing on offer, the waiting time at the checkout, etc. may play a role. Therefore, multiple regression is a much more appropriate technique to use to find out the effects of all the causal variables on the dependent variable. A typical research question would be:

Q7. To what extent do the quality of the merchandise, the waiting time at the checkout and the background music contribute to perceived customer satisfaction towards CTRL?

Chapter 7 is written by Siobhan Mc Carthy and Helen Xiaohong Chen and looks at logistic regression. In a multiple regression, the dependent variable is continuous. Sometimes, in a research scenario, the dependent variable can be binary or dichotomous. When this occurs, multiple regression is no longer useful and logistic regression is necessary to answer a research question such as:

Q8. To what extent do the quality of the merchandise, the waiting time at the checkout and the background music predict whether customers will shop in CTRL in the next week?

Daire Hooper wrote **Chapter 8** on exploratory factor analysis, which can be used to reduce data, to develop scales and to build theories. She used a dataset on service satisfaction and demonstrates how to run exploratory factor analysis and the reliability test.

Chapter 9 was written by Donncha Ryan on multidimensional scaling. This method is useful when the questionnaire is designed by using comparative scaling techniques, such as Q-sort or constant sum scaling, and the data needs to be explored to obtain some understandings for a researcher.

The datasets used in **Chapter 2**, **Chapters 3**, **4**, **5**, **6** and **7** and in **Chapter 8** are available for download at **www.oaktreepress.com**.

REFERENCES

Easterby-Smith, M., Thorpe, R. & Lowe, A. (2002). *Management Research: An Introduction*, 2nd ed. London: Sage Publications.

CHAPTER 2
SPSS FOR REAL BEGINNERS

Eddie Rohan

INTRODUCTION

Statistical Package for the Social Sciences (SPSS) is a commonly-used software package for quantitative investigations among social scientists. SPSS became PASW [3] for Version 17.0.3 released in September 2009 to Version 18.0.3 released in September 2010. The base SPSS software can perform the following statistics:

- Descriptive statistics (for example, frequencies, descriptive and descriptive ratio statistics);
- Bivariate statistics (for example, t-tests, ANOVA and correlation);
- Prediction for numerical outcome (for example, linear regression and logistic regression); and
- Prediction for identifying groups (for example, cluster analysis, factor analysis and discriminant scaling).

Following its acquisition by IBM, the name of the software was officially changed to IBM® SPSS in January 2010. Nevertheless, it is referred to in this book as SPSS, as this is how it is best-known still.

SCALES AND QUESTIONNAIRES

In any research project, it is essential that survey instruments or questionnaires are designed in a clear and unambiguous fashion, and that these surveys relate to a representative sample. Otherwise, there is no possibility of obtaining valid results in relation to the

---

[3]   The screen grabs are from PASW 18, the version of SPSS used by the authors.

target population. The size of a sample depends on the number of variables that are included in a research project. The complete questionnaire can be of any length or complexity, as long as it is well structured in sections and the length does not become an issue that discourages responses.

The eight most common questions used in business or social science research projects are illustrated in the sample questionnaire in **Figure 2.1.**

Questions 1, 2 and 3 deal with nominal scales or categorical scales where the researcher names or uses a category for the data in SPSS. For Question 1, '**Single**' can be coded as '**1**', '**Married**' as '**2**', '**Living as Married**' as '**3**', etc. The numbers do not mean anything apart from identifying the various marital states.

In Questions 2 and 3, only two mutually exclusive choices are presented and so they are dichotomous questions. The coding of a dichotomous question can be '**1**' *versus* '**2**', or '**1**' *versus* '**0**', as long as they are defined as nominal in data coding. Note that, while the films comprise one question in the questionnaire, the viewing of each film is a separate variable in SPSS.

Questions 4 and 5 deal with one of the most commonly-asked questions in a survey: the age of the respondents. However, both question formats have been criticised. Question 4 can discourage respondents from completing the questionnaire, as not many people like to disclose their age directly. The age ranges in Question 5 are those commonly reported by organisations such as the Central Statistics Office in Ireland. However, this format has been criticised by social scientists: the ranges may be offensive, arbitrary and limit the analysis (Gendall & Healy, 2008). For instance, in the Sample Questionnaire, a 65-year-old may differ from a 70-year-old, an 80-year-old or a 90-year-old, all of whom would be included in the one range. The most appropriate approach when using a Question 4 format is to make sure it does not become an issue. When using a Question 5 format, make sure the ranges are well justified and designed and, if the results will be compared with a previous study, ensure the age ranges are comparable. Alternatively, follow the advice provided in Gendall & Healy (2008) by asking for the respondents' year of birth.

Figure 2.1: Sample Questionnaire

Question 1   Marital Status (Please circle as appropriate)

Single	....	1
Married	....	2
Living as married	....	3
Separated/divorced	....	4
Widowed	....	5

Question 2   Gender (Please tick)

Female   ☐

Male   ☐

Question 3   Have you seen the following films? (Please circle as appropriate)

Pirates of the Caribbean	Yes	No
Happy Feet	Yes	No
Casablanca	Yes	No

Question 4   What was your age on your last birthday?

_____ years old.

Question 5   Your age (Please circle as appropriate)

Under 18	....	1
18-24	....	2
25-34	....	3
35-44	....	4
45-54	....	5
55-64	....	6
65 and over	....	7

Question 6   Please state your opinion regarding the following statement: *"A night out at the cinema is good value for money."*

Strongly Disagree	Disagree	Neutral	Agree	Strongly Agree
1	2	3	4	5

Question 7 Please state your opinion of Brand X

Boring	☐	☐	☐	☐	☐ ☐	☐	Exciting
Unreliable	☐	☐	☐	☐	☐ ☐	☐	Reliable
Inexpensive	☐	☐	☐	☐	☐ ☐	☐	Expensive

Question 8   Rank your choices when viewing films in order of preference with 1 being your first choice and 5 least favoured.

Go to the cinema	____
Rent a DVD	____
Download to computer from the web	____
View live on TV	____
Record from TV and view later	____

Questions 6 and 7 deal with interval data.[4] Question 6 is a 5-point Likert scale, while Question 7 is an example of a 7-point semantic differential. These are commonly used in social and marketing research.

Question 8 deals with ordinal data where, instead of selecting their favourite movie, respondents are asked to rank their choices *in order*. Therefore, evidence is obtained on their second, third and fourth choices as well as the respondent's first choice.

Sales figures, costs or the number of customers are ratio data, which are not included in this questionnaire. Also note that the classification section is best located at the end of interview questionnaires.

BASIC STEPS IN SPSS

Questionnaires can be administered by an interviewer or by means of phone, mail, email and online websites.

Once data is collected, the next step is to submit it to SPSS. There are three basic steps to create an SPSS file for further analysis:

- ♦ **Step 1:** Define variables;
- ♦ **Step 2:** Data input;
- ♦ **Step 3:** Data cleaning.

Step 1: Define Variables

When you open a new window in SPSS, a smaller window pops up asking '**What would you like to do?**'. Click on '**Type in data**' and then on '**OK**'. You will be brought to the window shown in **Figure 2.2**, which shows a grid that looks like a spreadsheet with rows and columns. The columns (across the top) are used for the variables (which are the questions or parts of questions), while the numbers down the left-hand side identify the survey respondents or the observations.

---

[4] There is a debate in the literature regarding whether Likert scales and semantic differential scales are interval or ordinal. The conditions for them being interval are that they are symmetric, equidistant and there is an arbitrary zero point. In this book, all Likert scales and semantic differential scales are treated as interval.

Figure 2.2: 'Main' Menu in SPSS on the 'Data View' Window (1)

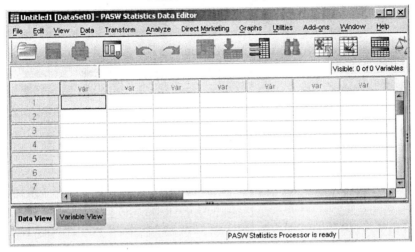

To define variables, click on '**Variable View**' at the bottom-left of the window. You will be brought to the following window to define variables, as shown in **Figure 2.3**.

Figure 2.3: 'Main' Menu in SPSS on the 'Variable View' Window (1)

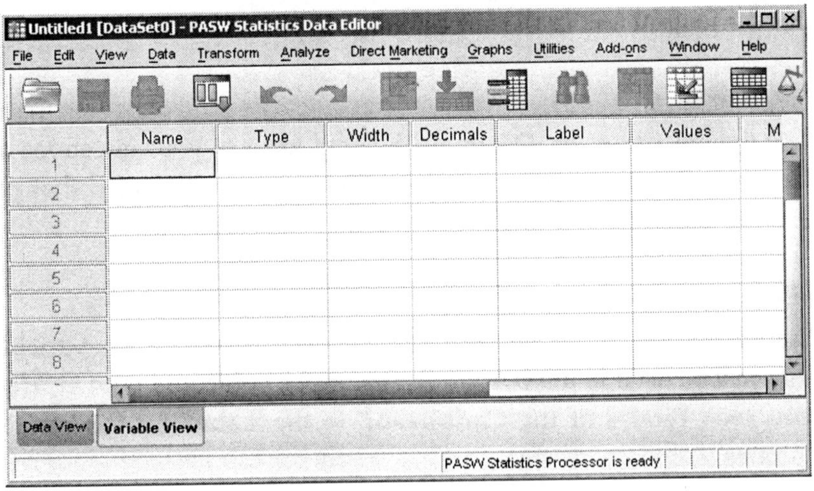

Name

To name a variable is the first job: each variable name must be unique and easy for you to connect to the question you have used in the questionnaire. The name must start with a letter and cannot contain spaces, numbers or symbols. For the Sample Questionnaire, the first variable is named '**marital_status**', the second '**gender**', the third '**pirates**', the fourth '**happy_feet**', etc. Note, both '**marital**' and '**status**' (and '**happy**' and '**feet**') must be linked by an underscore '_' as shown.

Type

The default value for '**Type**' is '**Numeric**', which is useful for most data. But if you wish to type in text information, you need to change '**Type**' from '**Numeric**' to '**String**'. For example, if you choose not to code '**marital_status**' in the Sample Questionnaire, click on '**Type**' and choose '**String**' so that you can type in the actual words the respondents provided to describe their marital status.

Width

The default value for '**Width**' is '**8**'. This is normally a good option. However, if you expect to type in text information or have very large values that will take up more than eight digits, you should set it as per the longest text or the largest value in your data.

Decimals

The default value for '**Decimals**' is '**2**'. If most of your questions are like those in the Sample Questionnaire, it is recommended that you set it at '**0**'.

Label

'**Label**' allows you to provide a longer description for the variables that will be used in the output. For example, you could type in '**Have you seen Pirates of the Caribbean?**' in the '**Label**' for the variable '**pirates**'.

Values

You must define the meaning of the values that you are going to use to code your variables in SPSS. For the variable **'marital_status'**, for example, define **'1'** for **'Single'**, **'2'** for **'Married'**, **'3'** for **'Living as Married'**, and so on. To do this, click on **'Values'**, and in the window as shown in **Figure 2.4**, type in **'1'** in the box for **'Value'**, and then **'Single'** in the box for **'Label'**. Once you click on **'Add'**, the value and label will be recorded in the big box next to **'Add'**. You need to do the same for **'2'** – **'Married'**; **'3'** – **'Living as Married'** and so on.

Figure 2.4: 'Value Labels' Dialogue Box

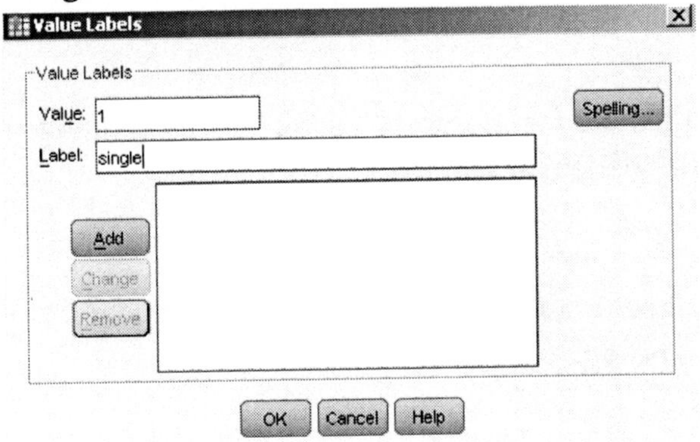

Missing

When you come across a blank answer to a question, a missing value occurs. To ask SPSS to identify a missing value easily, it is important that you define a missing value – for example, by using **'99'** for it. Thus, you need to click on **'Missing'** and set **'99'** to represent a missing value, as shown in **Figure 2.5**. Then when you are at the data input stage, simply type in **'99'** for any missing value you come across on the questionnaire.

Figure 2.5: 'Missing Values' Dialogue Box

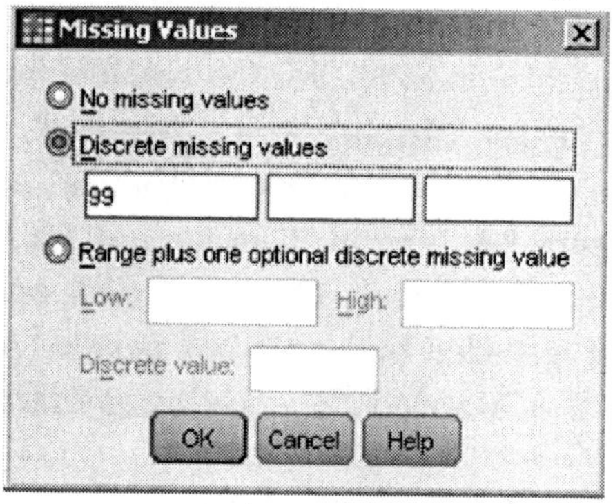

Columns

The default value for '**Columns**' is '**8**'. You may change it according to the values of data or names of variables.

Align

The default setting for '**Align**' is '**Right**'. Normally this is good and you do not need to change it.

Measure

This column refers to the measurement of each question – whether it is nominal, ordinal or interval (scale). Defining the '**Measure**' of each variable can smooth the way for further analyses in SPSS.

Once all variables are defined, your SPSS file on the '**Variable View**' window should look like the one in **Figure 2.6**.

Figure 2.6: 'Main' Menu in SPSS on the 'Variable View' Window (2)

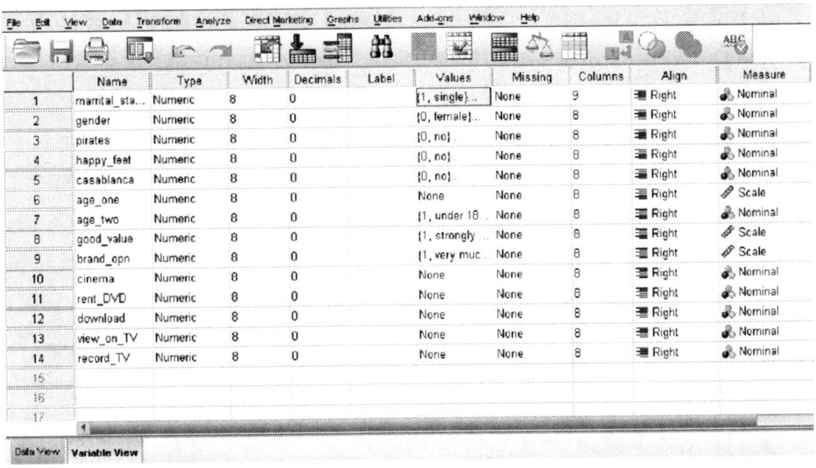

	Name	Type	Width	Decimals	Label	Values	Missing	Columns	Align	Measure
1	marital_sta...	Numeric	8	0		{1, single}...	None	9	Right	Nominal
2	gender	Numeric	8	0		{0, female}...	None	8	Right	Nominal
3	pirates	Numeric	8	0		{0, no}...	None	8	Right	Nominal
4	happy_feet	Numeric	8	0		{0, no}...	None	8	Right	Nominal
5	casablanca	Numeric	8	0		{0, no}...	None	8	Right	Nominal
6	age_one	Numeric	8	0		None	None	8	Right	Scale
7	age_two	Numeric	8	0		{1, under 18...	None	8	Right	Nominal
8	good_value	Numeric	8	0		{1, strongly ...	None	8	Right	Scale
9	brand_opn	Numeric	8	0		{1, very muc...	None	8	Right	Scale
10	cinema	Numeric	8	0		None	None	8	Right	Nominal
11	rent_DVD	Numeric	8	0		None	None	8	Right	Nominal
12	download	Numeric	8	0		None	None	8	Right	Nominal
13	view_on_TV	Numeric	8	0		None	None	8	Right	Nominal
14	record_TV	Numeric	8	0		None	None	8	Right	Nominal
15										
16										
17										

Step 2: Data Input

After defining all of the variables in your questionnaire, you now need to input all the answers selected by your respondents. You need to click on '**Data View**' to get to the window that allows you to enter data. Across a particular row on your worksheet, you should enter all the coded answers for one individual respondent. The result should look like the screen in **Figure 2.7**.

Figure 2.7: 'Main' Menu in SPSS on the 'Data View' Window (2)

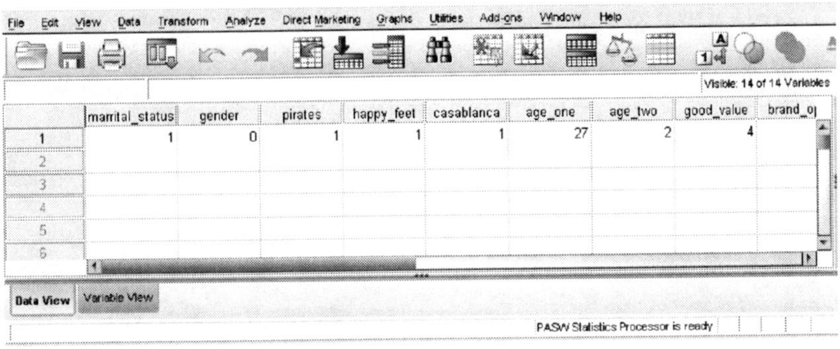

	marital_status	gender	pirates	happy_feet	casablanca	age_one	age_two	good_value	brand_o
1	1	0	1	1	1	27	2	4	
2									
3									
4									
5									
6									

Repeat this for each respondent until all the data has been input into SPSS. During data input, you should number each questionnaire with the row number on which it was input so that later you can trace a particular questionnaire that causes problems.

When you do not have access to SPSS, you can input your data in Microsoft Excel and then when you have access to SPSS later, you can import the Excel file into SPSS. For a detailed explanation of the procedure for doing this, refer to Pallant (2010). The important thing to remember is that, in Excel, you can have only 256 columns, which restricts the number of your variables to 256. If the number of variables in your data is more than that, you should be using SPSS for data input from the beginning.

Step 3: Data Cleaning

Typos

Typos – misspellings, transpositions of characters and digits, etc – are possible at the data inputting stage, so you always should double-check the data. As a first step in SPSS, use the frequency of each variable to identify typos. **Figure 2.8** illustrates how to perform frequency analysis.

Figure 2.8: 'Analyze' Menu in SPSS with 'Descriptive Statistics', 'Frequencies' and 'Descriptives' Submenus

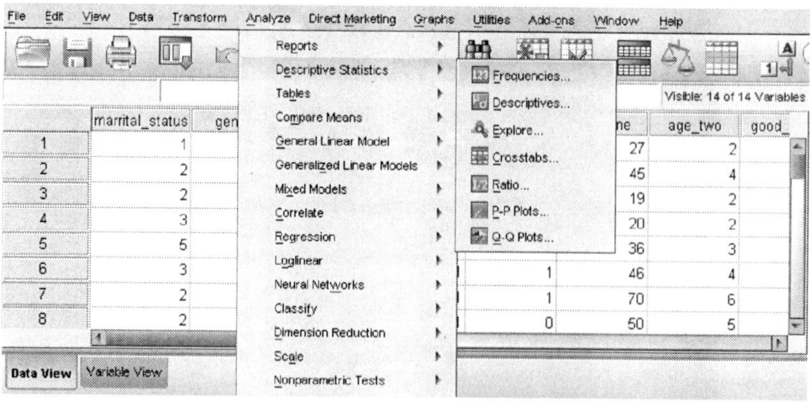

When you have created the Output file, you need to pay attention to any 'abnormal' values. For example, if you define **'gender'** as 'female' – '0' and 'male' – '1', then 'gender' '6' is an impossible value, as is 'gender' '10'. Similarly, Likert or semantic differential scales normally contain 1 to 5 or 1 to 7 intervals. If a value of '22' appears, that means something went wrong during input. You can trace the questionnaire by the number you wrote down during data input to find out whether the problem is due to a typo.

Missing Values

Missing values is another common problem in a dataset. You need to run '**Frequencies**' to get an idea of the percentage of values that are missing for each of the variables.

It is essential to understand why missing values happen. If it happens randomly and rarely, you probably can ignore it. But if it happens systematically, it may indicate that some instruments you have used in the questionnaire might not be appropriate.

There are three common methods in SPSS to deal with missing values that occur randomly under '**Option**' in '**Analysis**', such as regression, t-test, etc:

◆ If one case contains a missing value but has valid and necessary information on other variables, it is recommended that '**Exclude cases pairwise**' is selected. This allows the case to be included in other analyses.

◆ If '**Exclude cases listwise**' is chosen, then the whole case will be excluded for further analysis even if it carries values for other analyses.

◆ '**Replace with mean**' is recommended when the number of cases that contain missing values is small compared to the sample size.

MISCELLANEOUS

Four other basic tools merit mention: the Recode, Compute, Select Cases and Split File commands.

Recode

This is very important and should be known by all researchers. As you transform the data, you can recode the old variable into a different variable, or into the same variable.

Three research situations make this command very useful:

♦ Collapsing categories;

♦ Interval data into nominal data; and

♦ Reversing coding.

Collapsing Categories

Collapsing categories is where many values for each variable and a small sample size may cause problems in interpretation. Often, collapsing a large number of categories into fewer categories may have the effect of making things more manageable.

For example, in the Sample Questionnaire, the variable **'good_value'** ("A night out at the cinema is very good value for money") has five categories: **'strongly disagree'**, **'disagree'**, **'neither agree or disagree'**, **'agree'** and **'strongly agree'**. For simplicity and clarity, these might be collapsed into three categories: **'disagree overall'**, which includes values **'1'** and **'2'** in the dataset; **'neither agree or disagree'**, which includes the value **'3'** in the dataset; and **'agree overall'**, which includes values **'4'** and **'5'** in the dataset.

The new variables can be created by clicking on **'Transform'** and then **'Recode into Different Variables'**. You must type a name for the new variable – in this case, **'good_value_two'** – into **'Name'** under **'Output Variable'**, as shown in **Figure 2.9**. Once you click on **'Change'**, the new name will be created. You may leave **'Label'** blank if the name is clear to you.

Figure 2.9: 'Recode into Different Variables' Dialogue Box

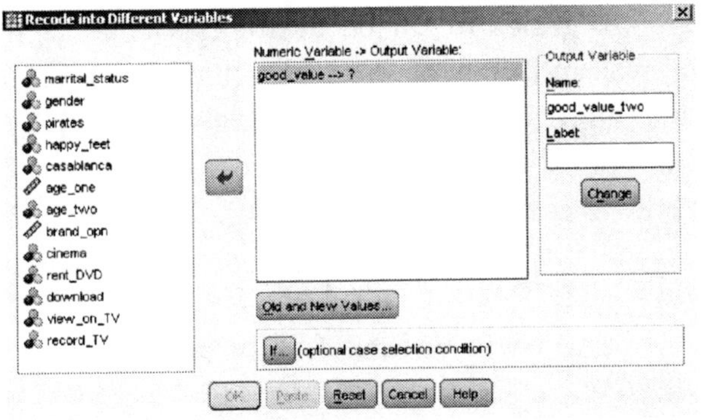

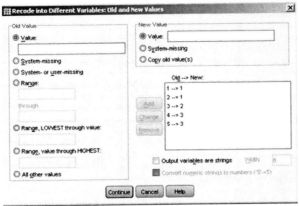

Then you need to click on 'Old and New Values' and define 'Old Value' and 'New Value'. In this case, you type '1' in 'Old Value' and code it to '1' in 'New Value'; '2' in 'Old Value' to '1' in 'New Value', '3' in 'Old Value' to '2' in 'New Value'; '4' in 'Old Value' to '3' in 'New Value'; '5' in 'Old Value' to '3' in 'New Value', as shown in Figure 2.10. You click on 'Continue' to create the new variable.

Figure 2.10: 'Recode into Different Variables: Old and New Values' Dialogue Box (1)

Interval Data to Nominal Data

Another use of recoding variables is to code interval data into nominal data. For example, in the Sample Questionnaire, you can transform 'age_one' into nominal data, where either two or three suitable age categories might be useful classifiers of respondents. Generally, such recoding of interval values (such as age) can be classified into high, middle and low (or high and low) categories that are suitable for analyses such as cross-tabulation. **Figure 2.11** illustrates how to perform such a recoding.

Figure 2.11: 'Recode into Different Variables: Old and New Values' Dialogue Box (2)

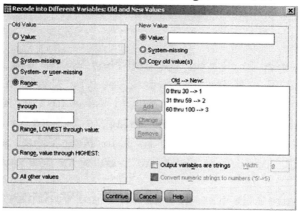

Reverse Coding

While attitude questions are best presented in a balanced way, having both positive and negative wording, the question of consistency of meaning is an important issue to be addressed. "**A night out at the cinema is very good value for money**" is an example of positive wording, where '**Strongly disagree**' might have a score of '**1**' and '**Strongly agree**' a score of '**5**'. In such a case, the higher the score, the more favourable the opinion that cinema is good value for a night out.

On the other hand, if the question had been worded negatively as "**A night out at the cinema is not very good value for money**", then the use of '**Strongly disagree**' with a score of '**1**' and '**Strongly agree**' scoring '**5**' should be reverse-scored to ensure a consistent meaning.

Once you click on 'Recode into the same variable', you need to reverse-code '5' in 'Old Value' to '1' in 'New Value'; '4' in 'Old

Value' to '2' in 'New Value', '3' in 'Old Value' to '3' in 'New Value'; '2' in 'Old Value' to '4' in 'New Value'; and finally '1' in 'Old Value' to '5' in 'New Value'. Such a transformation provides the desired consistency of meaning.

Compute

New variables can be generated by using the existing variables to create useful data. For example, it looks useful in the Sample Questionnaire to count the number of films that each person has seen. Movie-lovers who have seen everything then can be distinguished from people who rarely watch a movie. Determining the profile of heavy users is necessary for most businesses, given the relative importance of those customers.

The response '1' in the Sample Questionnaire for Question 3 indicates that a respondent has seen a film such as 'Casablanca', while '0' means that they have not. Looking across the file, the addition of scores relating to the three films gives the total number of films seen. Once you click on 'Compute Variable' under 'Transform', you will be brought to the window shown in **Figure 2.12**.

Figure 2.12: 'Compute Variable' Dialogue Box

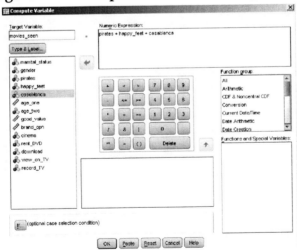

You will need to type '**movies_seen**' into '**Target Variable**'. As the '**=**' symbol indicates, you need to work out your equation. In this case, you simply want to add the responses to the three movies to understand how much the respondents like to watch movies – thus: '**pirates + happy_feet + casablanca**'. Clicking on '**OK**' creates the new variable.

Select Cases

Sections of an overall sample can be examined separately. For example, you can pick out respondents aged 45 and over (including 45-year-olds) from the sample to examine their behaviour. This selection of the data is found under the '**Data**' menu.

Once you are in the window as shown in **Figure 2.13**, choose '**age_one**' in this case. Click on '**If condition is satisfied**'. Then click on '**If**'.

Figure 2.13: 'Select Cases' Dialogue Box

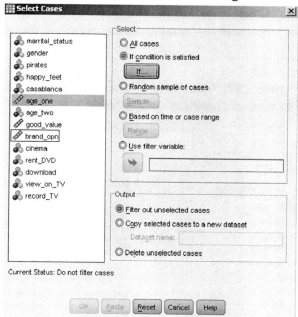

Then define what you are looking for: cases where the respondents are 45 years old or greater (age >= 45), as shown in **Figure 2.14**.

Figure 2.14: 'Select Cases: If' Dialogue Box

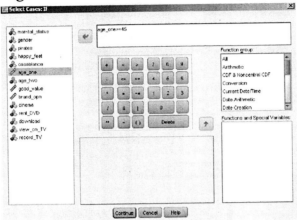

Split Files

Finally, you can split the file to obtain, for example, the breakdown of the dataset by age, profession, gender, etc. If you want to obtain only the female respondents' data, you can click on 'Data' and then click on 'Split Files'. You will be brought to the window shown in Figure 2.15.

Click on 'Compare groups', and then choose 'gender' and drag it into the 'Groups Based on' box. Once you click on 'OK', the data will be rearranged by gender, all female respondents' data being grouped together and all male respondents' data being grouped together.

Figure 2.15: 'Split File' Dialogue Box

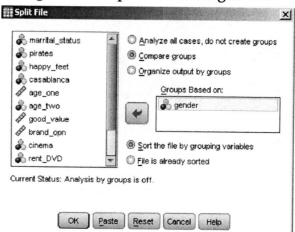

Select Cases and Split File will remain operational until you return to these commands and tick the box **"Use ALL cases"**.

Finally, save your file to have it ready for further analyses.

REFERENCES

Gendall, P. & Healey, B. (2008). 'Asking the Age Question in Mail and Internet Surveys', *International Journal of Marketing Research*, 50(3): 309-316.

Pallant, J. (2010). *SPSS Survival Manual*. New York: McGraw Hill.

CHAPTER 3
PRELIMINARY DATA ANALYSIS

Helen Xiaohong Chen

Once you have created the SPSS file, the next step is to run some preliminary or exploratory analyses. The benefits that can be achieved by doing this include:

- Gaining a basic understanding of the characteristics of the dataset;
- Identifying outliers and other possible errors;
- Examining whether the variables are normally distributed (in terms of univariate and multivariate normality) if normal distribution is a requirement for further data analysis and so on.

This preliminary data analysis is so important that you cannot afford to omit it before running your selected analyses such as t-test, MANOVA, regression, etc.

This chapter demonstrates how the preliminary analyses were carried out for the analyses discussed in **Chapters 4, 5, 6 and 7**, without focusing on the statistical theories behind the techniques. Interested readers may refer to Tabachnick & Fidell (2001) and/or Pallant (2007).

DESCRIPTIVES

Continuous Variables

Descriptives usually are run to help a researcher to understand the basic features of the dataset and to examine the normality of variables. The following steps demonstrate how to run descriptive statistics for continuous variables.

Step 1

Click on '**Analyze**' and then '**Descriptive Statistics**' and then '**Descriptives**', as shown in **Figure 3.1**.

Figure 3.1: 'Analyze' Menu in SPSS with 'Descriptive Statistics' and 'Descriptives' Submenus

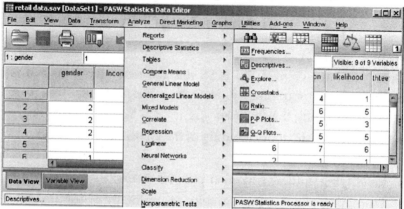

Step 2

Move all the continuous variables into '**Variable(s)**' and then click on '**Options**' as shown in **Figure 3.2**.

Figure 3.2: 'Descriptives' Dialogue Box

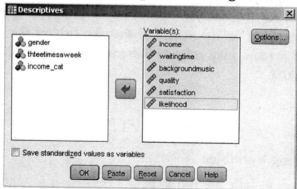

Step 3

In the '**Options**' window as shown in **Figure 3.3**, ensure the following statistics are ticked: '**Mean**', '**Std. deviation**', '**Minimum**', '**Maximum**', '**Kurtosis**', and '**Skewness**'. Then click on '**Continue**' and then '**OK**'.

Figure 3.3: 'Descriptives: Options' Dialogue Box

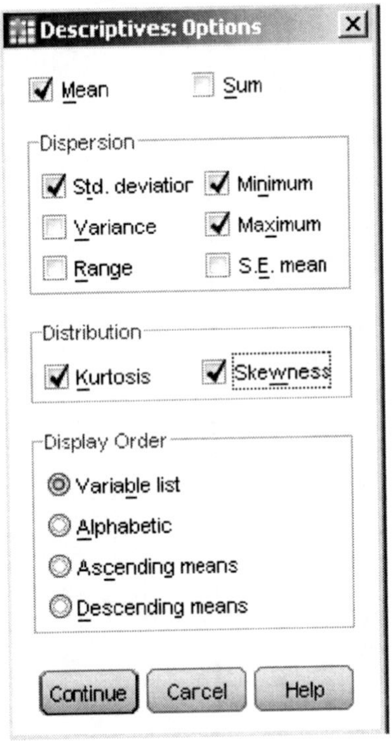

Table 3.1 presents the descriptive statistics output.

It shows the **minimum** and **maximum**, which can be used to detect typos in data input, as discussed in **Chapter 2**. The mean and standard deviation also are obtained.

Skewness is used to understand whether the data is symmetric. Therefore, for a perfectly normally distributed dataset, the skewness is zero. A negative value for skewness indicates that the left tail of the bell shape of the distribution of the data is longer than the right tail – the data is skewed left.

Table 3.1: Descriptive Statistics

| | N | Minimum | Maximum | Mean | Std. Deviation | Skewness | | Kurtosis | |
	Statistic	Statistic	Statistic	Statistic	Statistic	Statistic	Std. Error	Statistic	Std. Error
waitingtime	80	1	7	4.63	1.554	-.364	.269	-.443	.532
backgdmusic	80	1	6	2.21	1.290	1.480	.269	1.798	.532
quality	80	2	7	4.19	1.631	.120	.269	-1.137	.532
satisfaction	80	1	7	3.95	1.630	.208	.269	-.668	.532
likelihood	80	1	7	3.81	1.677	.106	.269	-.647	.532
Valid N (listwise)	80								

Since perfectly normal distribution is rare, use the following technique to accept/reject whether a variable is normally distributed by skewness (Tabachnick & Fidell, 1996). First, multiply the standard error of skewness of '**waitingtime**' by two: 0.269 x 2 = 0.538. Then examine whether the skewness statistic for '**waitingtime**' falls within the range -0.538 to +0.538. In this example, all variables except one have skewness values within the range – thus '**backgdmusic**' is not normally distributed by skewness.

Kurtosis also can be used to examine whether a dataset is normally distributed. Kurtosis refers to the degree of peakedness of a distribution. A normal distribution needs the Kurtosis value to be close to zero. A positive Kurtosis indicates 'peaked' distribution of the dataset, while a negative Kurtosis indicates a 'flat' distribution of the dataset. Follow the same method as in skewness and then examine whether the Kurtosis values fall into the ranges. In this example, '**quality**' and '**backgdmusic**' are not normally distributed by Kurtosis.

Using **Skewness** or **Kurtosis** to examine the normality of variables is a basic data analysis method. In this example, you need to ensure that the two dependent variables '**satisfaction**' and '**likelihood**' are normally distributed. More advanced techniques – such as the Kolmogorov-Smirnov (K-S) and Shapiro-Wilk (S-W) tests, or the normality plots of the data (normal Q-Q plots) – are available to achieve the same objective.

Categorical Variables

You should use frequency to obtain descriptive statistics for categorical data. This will allow you to understand the characteristics of the respondents, such as gender and age, as well as to obtain qualitative data such as movies that respondents have seen (see **Chapter 2**). To use frequencies:

- Click on '**Analyze**' and then '**Descriptive Statistics**' and then '**Frequencies**'.
- Choose the categorical variables you want to analyse and move them into the '**Variables**' box.
- Click on '**Continue**' and then '**OK**'.

GRAPHS

Graphs provide an alternative approach to understanding the features of the dataset, such as normal distribution and homoscedasticity. SPSS can generate a number of different types of graph: three will be discussed in this chapter.

Scatterplots

Scatterplots can be used to explore the relationship between two continuous variables. Before you run correlations, you are recommended to generate scatterplots to understand whether there is a linear relationship between any two variables.

The following steps demonstrate how to generate scatterplots.

Step 1

Click on '**G**raphs', then on '**L**egacy Dialogs' and then on '**S**catterplots' as shown in **Figure 3.4**.

Figure 3.4: 'Graphs' Menu in SPSS with 'Legacy Dialogs' and 'Scatter/Dot' Submenus

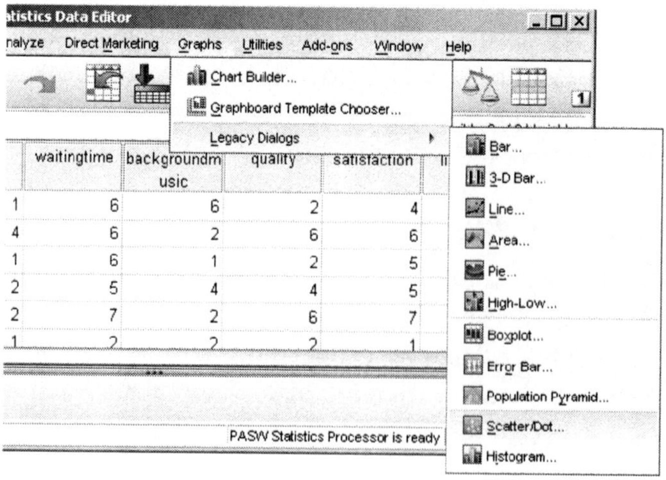

Step 2

When a window like the one in **Figure 3.5** pops up, click on '**Simple Scatter**'. Then click on '**Define**'.

Figure 3.5: 'Scatter/Dot' Dialogue Box

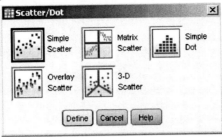

Step 3

On the next window as shown in **Figure 3.6**, move one variable, such as '**satisfaction**', into the box under '**Y Axis**' and another variable, such as '**waitingtime**', into the box under '**X Axis**'. To examine the relationship between them, move '**ID**'[5] into the box '**Label Cases by**'. You can name the scatterplots by clicking on '**Titles**'.

Figure 3.6: 'Simple Scatterplot' Dialogue Box

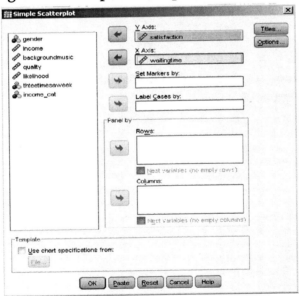

---

[5]   This allows you to know which cases are outliers should any outliers be identified.

Step 4

Clicking on 'OK' generates the scatterplot as shown in **Figure 3.7**.

Figure 3.7: Scatterplot for 'satisfaction' and 'waitingtime'

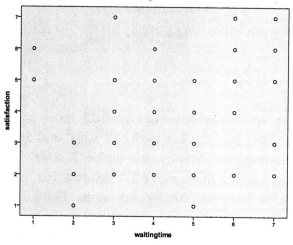

Step 5

Double-click on **Figure 3.7** in the output file to be brought to the 'Chart Editor'. Click on 'Add Fit Line at Total' as indicated by the circle in **Figure 3.8**.

Figure 3.8: Chart Editor for Scatterplots

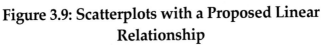

This will give a proposed linear relationship line between the two variables with R^2 indicated on the top right (see **Figure 3.9**).

Figure 3.9: Scatterplots with a Proposed Linear Relationship

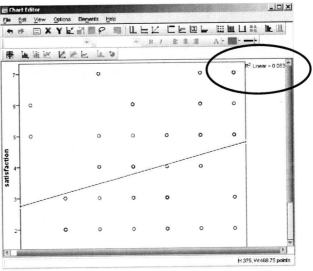

Scatterplots help you to understand characteristics of the data, such as normal distribution and homoscedasticity.

If the scatterplot generated in SPSS looks like the one depicted in **Figure 3.10**, then there are no major issues with the data. However, if the scatterplot looks like any of those in **Figure 3.11** or **Figure 3.12**, then your data has heteroscedasticity problems.

Figure 3.10: Homoscedasticity (stability of variance is evident)

Figure 3.11: Heteroscedasticity (variance reduces over different levels of another variable)

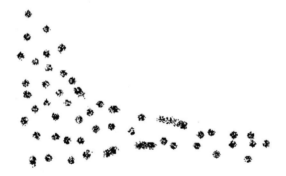

Figure 3.12: Heteroscedasticity (variance changes over different levels of another variable)

Histograms and Normal Q-Q Plots

You can use histograms and normal Q-Q plots to examine the distribution of a continuous variable. The following steps demonstrate how to generate histograms and normal Q-Q plots.

Step 1

Click on '**Analyze**', then on '**Descriptive Statistics**', then on '**Explore**' as shown in **Figure 3.13**.

Figure 3.13: 'Analyze' Menu in SPSS with 'Descriptives Statistics' and 'Explore' Submenus

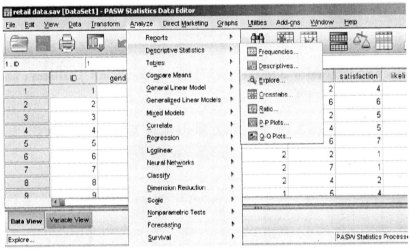

Step 2

In the window as shown in **Figure 3.14**, move the continuous variable that you want to test for normality into the box '**Dependent List**' and move '**ID**' into the box '**Label Cases by**'.

Figure 3.14: 'Explore' Dialogue Box

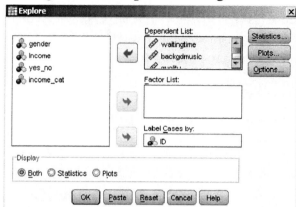

Step 3

Click on '**Statistics**' and then tick '**Descriptives**' and '**Outliers**'. Then click on '**Continue**'.

Figure 3.15: 'Explore: Statistics' Dialogue Box

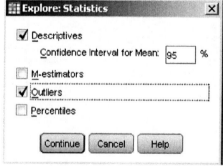

Step 4

Click on '**Plots**' in **Figure 3.14** and then tick '**Histogram**' and '**Normality Plots with test**' as shown in **Figure 3.16**.

Figure 3.16: 'Explore: Plots' Dialogue Box

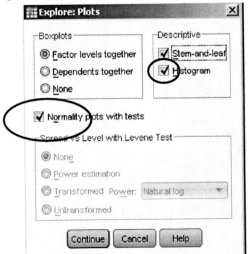

Step 5

Click on '**Continue**' and then '**OK**'.

The following (selected) histograms and normal Q-Q plots will be displayed in the output.

Figure 3.17: Histogram for 'waitingtime'

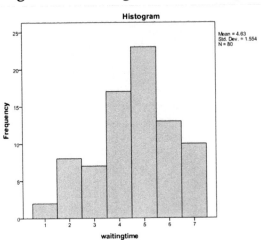

Figure 3.18: Histogram for 'backgdmusic'

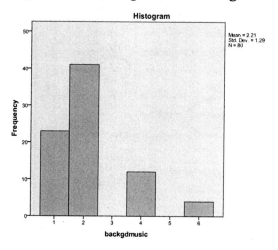

Figure 3.19: Normal Q-Q Plots for 'satisfaction'

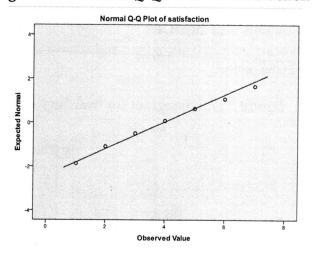

Figure 3.20: Normal Q-Q Plot for 'likelihood'

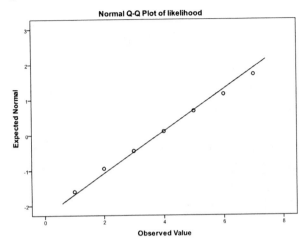

Figure 3.21: Normal Q-Q Plot for 'quality'

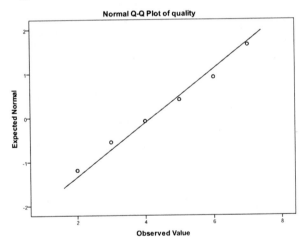

CORRELATIONS BETWEEN VARIABLES

Correlations between variables are important for particular types of analyses – for example, in MANOVA, the dependent variables need to be moderately correlated. However, in a multiple regression model, high correlations among the independent variables create a problem of multicollinearity. Therefore, the data needs to be carefully investigated by using correlations.

To run a test on correlations between variables, follow these steps.

Step 1

Click on '**A**nalyze', then on '**C**orrelate' and then on '**B**ivariate'.

Step 2

Move the continuous variables reflected in the Sample Questionnaire into the box '**V**ariables'. Tick '**P**earson' if the variables are interval, as shown in **Figure 3.22**. However, if the variables are ratio and ordinal data, tick '**S**pearman' correlation coefficients.

Step 3

Click on '**OK**' to get the results shown in **Table 3.2**.

Figure 3.22: 'Bivariate Correlations' Dialogue Box

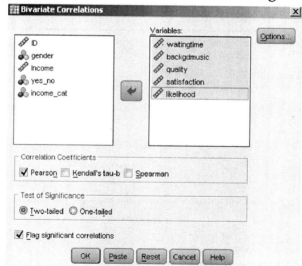

It is recommended that the greyed numbers in **Table 3.2** are not presented in the final table as these results are repeats.

Table 3.2: Correlation Coefficients between Variables

		waitingtime	backgdmusic	quality	satisfaction	likelihood
waitingtime	Pearson Correlation	1	.078	.003	.287**	.123
	Sig. (2-tailed)		.491	.978	.010	.276
	N	80	80	80	80	80
backgdmusic	Pearson Correlation	.078	1	-.176	.150	.118
	Sig. (2-tailed)	.491		.119	.185	.297
	N	80	80	80	80	80
quality	Pearson Correlation	.003	-.176	1	.308**	.374**
	Sig. (2-tailed)	.978	.119		.005	.001
	N	80	80	80	80	80
satisfaction	Pearson Correlation	.287**	.150	.308**	1	.626**
	Sig. (2-tailed)	.010	.185	.005		.000
	N	80	80	80	80	80
likelihood	Pearson Correlation	.123	.118	.374**	.626**	1
	Sig. (2-tailed)	.276	.297	.001	.000	
	N	80	80	80	80	80

** Correlation is significant at the 0.01 level (2-tailed).

Table 3.2 shows that no correlation coefficient between variables is greater than 0.9, eliminating the problem of collinearity. In **Chapter 5** on MANOVA, there are two dependent variables, '**satisfaction**' and '**likelihood**'. One prerequisite is that they must be moderately correlated. The correlation coefficient of 0.641 at a significance level of .000 reflects such a relationship and so you can safely proceed to MANOVA.

MULTIVARIATE OUTLIERS AND NORMALITY

Multivariate outliers, as compared to univariate outliers, refer to 'cases with an unusual combination of scores on two or more variables' (Tabachnick & Fidell, 2001: 67). Multivariate normality refers to a situation when the dependent variable is normally distributed with respect to each one of other variables, either dependent or independent. Multivariate normality takes a form of symmetric three-dimensional bells, specifically when the x-axis is the value of the dependent variable, the y-axis is the value of an independent variable, and the z-axis is the value of any other variable under consideration.

There are many methods to examine the multivariate normality, such as Mardia's statistic of multivariate skewness, Mardia's statistic of multivariate Kurtosis and the Doornik-Hansen multivariate normality test. However, the Mahalanobis distance is a popular method to calculate multivariate outliers (Tabachnick & Fidell, 2001) and multivariate normality (Pallant, 2007). It refers to the distance of a particular case from the centroid (the point created by the means of all variables) of the remaining cases (Tabachnick & Fidell, 2001).

The following procedure demonstrates how to generate Mahalanobis distance in SPSS and then how to use it to evaluate the multivariate normality.

Step 1

Click on '**A**nalyze', then on '**R**egression' and then on '**L**inear'.

Figure 3.23: 'Analyze' Menu in SPSS with 'Regression' and 'Linear' Submenus

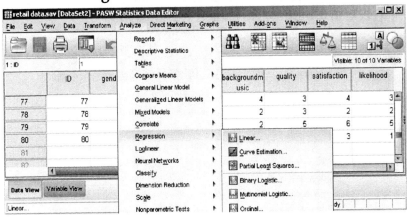

Step 2

Move the variable 'ID' into the box '**Dependent**' and the dependent variables that you are going investigate in the multivariate analyses into the box '**Independent(s)**' – in this case, '**satisfaction**' and 'likelihood', as shown in **Figure 3.24**.

Figure 3.24: 'Linear Regression' Dialogue Box

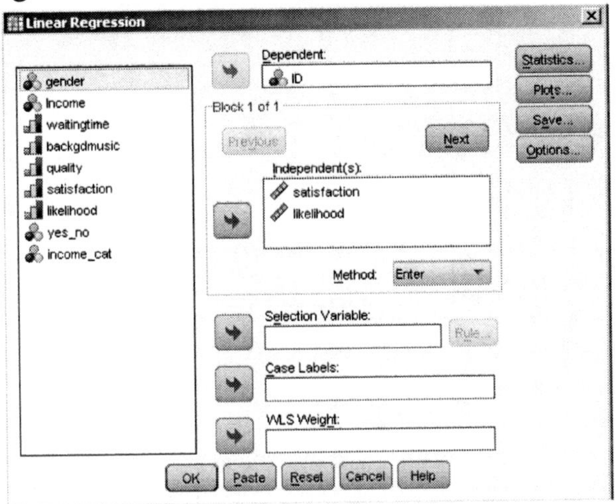

Step 3

Click on '**Save**' and then tick '**Mahalanobis**' under '**Distances**' and then click on '**Continue**', as shown in **Figure 3.25.**

Figure 3.25: 'Linear Regression: Save' Dialogue Box

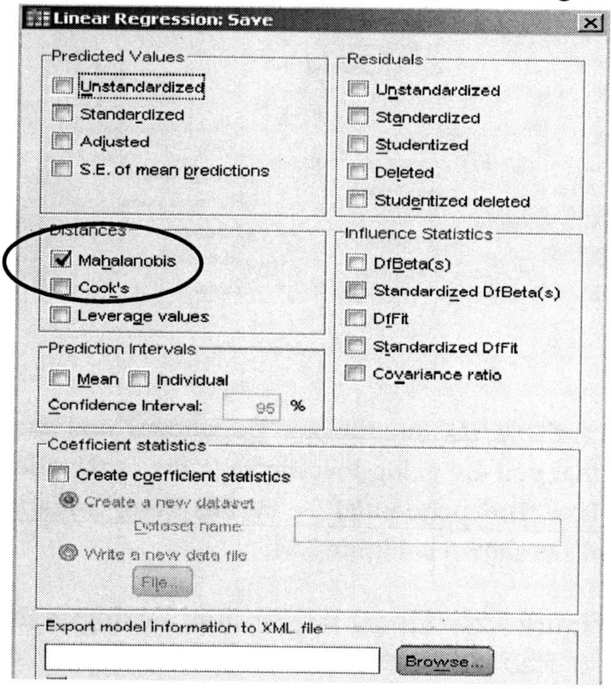

Table 3.3 presents some of the outputs.

Table 3.3: Critical Values for Evaluating Multivariate Normality

Number of dependent variables	Critical value
2	13.82
3	16.27
4	18.47
5	20.52
6	22.46

To examine the multivariate normality and outliers, you need to compare the Mahalanobis distance in **Table 3.3** against a critical value in **Table 3.4**, which is a Chi-square value that has been set by statisticians, such as Pearson & Hartley (1958) and Tabachnick & Fidell (1996) and summarised in Pallant (2007). In this example, the maximum Mahalanobis distance is 8.021 for two dependent variables, smaller than the critical value for two dependent variables (13.82). Therefore, you can conclude that the two dependent variables that will be included in MANOVA satisfy the requirement of multivariate normality.

Table 3.4: Residuals Statistics

	Minimum	Maximum	Mean	Std. Deviation	N
Predicted Value	35.50	45.07	40.50	2.099	80
Std. Predicted Value	-2.380	2.177	.000	1.000	80
Standard Error of Predicted Value	2.641	7.916	4.379	1.206	80
Adjusted Predicted Value	32.61	48.67	40.52	2.447	80
Residual	-38.192	41.099	.000	23.143	80
Std. Residual	-1.629	1.753	.000	.987	80
Stud. Residual	-1.699	1.783	.000	1.007	80
Deleted Residual	-41.667	42.487	-.022	24.100	80
Stud. Deleted Residual	-1.721	1.809	.000	1.013	80
Mahal. Distance	.015	8.021	1.975	1.631	80
Cook's Distance	.000	.091	.014	.018	80
Centered Leverage Value	.000	.102	.025	.021	80

REFERENCES

Pallant, J. (2007). *SPSS Survival Manual*, 3rd ed. Maidenhead: Open University Press.

Pearson, S. & Hartley, H. (1958). *Biometrika Tables for Statisticians*, Vol.1. Cambridge: Cambridge University Press.

Tabachnick, B.G. & Fidell, L.S. (1996). *Using Multivariate Statistics*, 3rd ed. New York: Allyn & Bacon.

Tabachnick, B.G. & Fidell, L.S. (2001). *Using Multivariate Statistics – International Students Edition*. New York: Allyn & Bacon.

CHAPTER 4
T-TESTS AND ANOVA

Agnes Maciocha

INTRODUCTION

The t-test and ANOVA (Analysis of Variance) are a family of parametric techniques that are used when analysing the differences between groups of subjects in terms of a particular feature. For example, you would use them if you wanted to compare the financial performance of service *versus* manufacturing firms or whether training has an impact on group performance.

The number of groups investigated influences the method to be used. Apply:

- T-test – if you want to analyse the differences between two groups or conditions;
- ANOVA – if you want to analyse whether there are differences between two or more groups or conditions.

Both t-test and ANOVA are concerned with the differences of means of groups (do not assume ANOVA is concerned with the differences between variances due to its name.).

DESCRIPTION OF THE TECHNIQUE

Conducting a comparison between two or more sample means enables researchers to generalise the difference between the respective population means (statistical inference). Thus, in your research, you should conduct a hypothesis test. Using the data from your sample, you would choose between a null and an alternative hypothesis. The null hypothesis (H_0) usually states there are no differences between population means, whereas the alternative hypothesis (H_1) claims that H_0 is not true.

T-test

The t-test assesses whether there are significant statistical differences of means between two groups.

There are two types of t-tests (see **Figure 4.1**):

♦ One sample t-test;
♦ Two sample t-test.

Figure 4.1: Different Types of t-tests

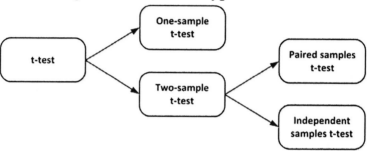

The one-sample t-test is used to compare the means of a sample to a known value. In most cases, the known value is the population mean. The two-sample t-test is based on the comparison of the means that are calculated using the data from two samples. It is possible to say that the t-test itself represents the distance between these two samples means (Sirkin, 2006). By comparing the value of your t-test to the t-critical value, you can check how likely it is that the distance between these two values differs by as much as you observe. In other words, you are checking how probable it is that this difference occurs only due to the sampling error.

In addition, as shown by **Figure 4.1**, there are two types of the two sample t-test statistic, with the choice of which to use depending on the type of the research design:

♦ **Paired samples t-test (dependent sample t-test):** Members of one sample are determined by the makeup of the other sample: you test the same participants on more than one occasion (Pallant, 2007);

♦ **Independent samples t-test**: Analysed samples are totally independent: you analyse differences between gender, people of different age, two different countries or political parties on one occasion.

When using the paired samples t-test, you are interested in the analysis of the differences between the means of two samples that are related – results of the first sample are dependent on the results of the second sample (repeated measurements, matched samples). Therefore, the paired sample t-test also is called the dependent t-test. Rather than focusing on the individual values, you analyse the difference between the two related values – before and after an experiment.

In using the independent samples t-test, your samples need to be unrelated as you analyse the means of two independent groups. However, you need to satisfy the assumption regarding equality of variances between the two samples. This is done by applying Levene's test of equality of variances (sometimes known as the F-test for homogeneity of variances). The results of this test determine the proper t-test for checking the means differences:

♦ If the significance of this test is greater than .05, then you *can* assume equal variances: ($\sigma_1^2 = \sigma_2^2$) and choose the pooled-variance t-test;

♦ If the significance of this test is smaller than .05, then you *cannot* assume equal variances ($\sigma_1^2 \neq \sigma_2^2$) and you need to choose the separate-variance t-test.

Finally, having calculated the value of the t-test, you can check whether the differences between your samples are significant. If the significance of your value is equal to or less than .05, then you can state that there is significant difference between the two groups. If the significance is higher than .05, then you accept H_0 and agree that there is no significant difference between the two groups.

ANOVA

If you have more than two groups to compare in a study, you need to use analysis of variance (ANOVA). Depending on the number of grouping variables, ANOVA can be classified as:

+ **One-way ANOVA**: There is only one grouping variable containing more than two groups – this type of ANOVA will be discussed in detail in the following section;
+ **Two-way ANOVA**: There are two grouping variables;
+ **Factorial ANOVA**: There are more than two grouping variables.

ANOVA constitutes an extension of the independent samples t-test assuming equal variances (Levine *et al.*, 2001). Consequently, like when using the t-test, you calculate some statistics – in this case, the F statistic. The F-test is a measure of the ratio between the variation explained by the model and the variation explained by unsystematic factors (Field, 2000). It is calculated as the ratio of the mean square between to the mean square within samples:

$$F = \frac{MS_B}{MS_W} = \frac{\text{variance between samples}}{\text{variance within samples}}$$

In order to conclude that there are significant differences between the groups (and thus reject H_0), the value of this statistic must be greater than 1. Consequently, the variance accounted by particular features of the groups analysed must be greater than the variance that existed within the samples analysed. To verify this, you need to look at the significance value of the F-test. If it is less than or equal to .05, you can conclude that, in the population, there is at least one inequality. However, you do not know which means (groups) are different. To determine that, SPSS offers two options:

+ **Planned contrasts** are performed if you have specific hypotheses that you want to test. Planned contrasts (*a priori*) are applied when there is a need to test specific hypotheses that, very often being described as a result of a theory or previous research, are planned before the analysis of the data. They are not recommended when there are a large number of differences to be analysed (Pallant, 2007);

♦ **Post-hoc tests** are performed when you have no specific hypotheses. They are used when there is a need to run out the whole set of comparisons. There are two steps to follow: (1) verify whether there are any significant differences among the groups (F-ratio needs to be significant); (2) perform a second test in order to recognise where these differences arise.

If the dependent variable is score or interval data, parametric ANOVA is used (as demonstrated in the following section). However, if the dependent variable is non-parametric in rank or order data, then you are recommended to use 'Non-parametric tests' under the '**Analyze**' submenu in the '**Main**' menu of SPSS.

DATA REQUIREMENTS AND ASSUMPTIONS

Type of Variables

In order to conduct t-tests or ANOVA, you need to have two types of variables:

♦ **One continuous dependent variable:** Financial performance, training results, etc.;

♦ **One categorical independent variable(s):** Gender, type of industry – sometimes, this can be a continuous variable and be converted into categorical data (see **Chapter 2** for details).

Homogeneity of Variance Assumption

ANOVA is fairly robust in terms of violations of the homogeneity of variance assumption when sample sizes are equal. This is not always true, however, when these sample sizes are not equal. In such situations, you cannot use the F-test – instead, you can use Welch's F (Field, 2000).

Normality of Distribution

In both t-tests and ANOVA, the samples need to be normally distributed. The normality of the distribution in turn depends on the sample size. If the sample size is bigger than 30,[6] you can assume normal distribution of that sample.

Errors and Power of the Test

Errors

T-tests and ANOVA are procedures that aim to test hypotheses. These approaches, therefore, include the possibility of a wrong conclusion being drawn.

You can make two types of error:

♦ **Type I error (alpha error):** The probability of falsely rejecting a true null hypothesis (H_0), where you state that there are differences between the groups investigated but, in reality, there are not.

♦ **Type II error (beta error):** The probability of failing to reject a false null hypothesis, where you conclude that the groups do not differ while, in reality, they do.

The Power of the Test

The test indicates the probability that you reject the null hypothesis when it is actually false. It is inversely related to the Type II error: as the chances of Type II error decrease, the power of the test increases. The power of the test is directly related to the sample size – with small samples (less than 20), there is a possibility that a non-significant result may happen because of the lack of sufficient power. With relatively big samples (more than 100), there should be no concerns about the power of the test (Stevens, 1996).

A WORKED EXAMPLE

Independent Samples T-test

The research objective: To investigate whether there are differences between female and male consumers regarding their perceived satisfaction level towards CTRL, a clothes retailer.

---

6   Many statisticians use this as a rule of thumb, but students are recommended to follow the techniques discussed in **Chapter 3** to examine normal distribution.

Independent variable: 'gender'.

Dependent variable: Perceived satisfaction towards CTRL, 'satisfaction'.

Step 1

In the menu '**A**nalyze', choose '**Co**mpare Means' and then 'Independen**t**-Samples T Test' (**Figure 4.2**).

Figure 4.2: 'Analyze' Menu in SPSS with 'Compare Means' and 'Independent-Samples T Test' Submenus

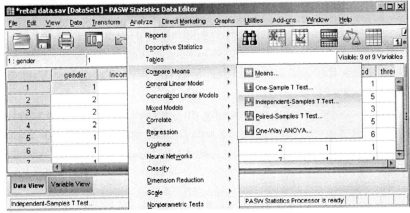

Step 2

In the next window that pops up, as shown in **Figure 4.3**, highlight the variable '**satisfaction**' and click on ➡ to place the variable 'satisfaction' in the box '**T**est Variable(s)'.

Figure 4.3: 'Independent-Samples T Test' Dialogue Box

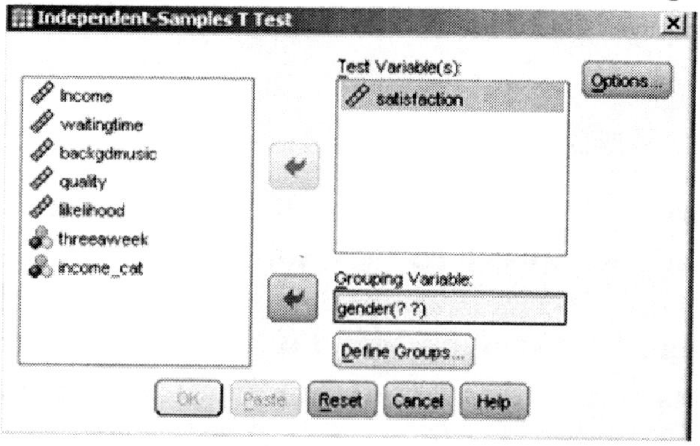

Step 3

Then place the variable '**gender**' into the box '**Grouping Variable**' (see **Figure 4.3**).

Then define groups by clicking on '**Define Groups**'. The window shown in **Figure 4.4** will pop up. In this case, use '**1**' for female customers, and '**2**' for male consumers. Click on '**Continue**' and then '**OK**'.

Figure 4.4: 'Define Groups' Dialogue Box

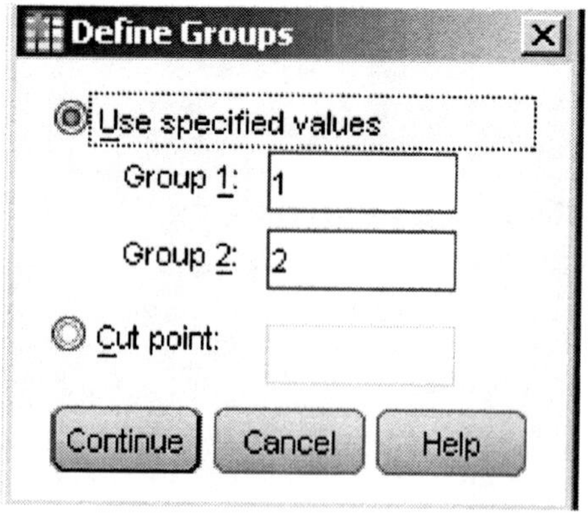

Tables 4.1 and **4.2** show the SPSS outputs.

Table 4.1: Group Statistics

	Gender	N	Mean	Std. Deviation	Std. Error Mean
satisfaction	Female	52	3.75	1.619	.225
	Male	28	4.32	1.611	.305

Interpretation

Table 4.1 presents descriptive statistics of the perceived satisfaction level of female and male customers towards CTRL. The sizes of these two groups are as follows: 52 female consumers and 28 male consumers (second column – depicted as 'N'). Note that the mean for the female consumers (3.75) is less than the mean for the male consumers (4.32).

To understand whether such a difference is statistically significant, you need to examine the actual t-tests presented in **Table 4.2**. First, you need to verify the assumption about equality of variances. To do this, look at the first column that provides the F-test under 'Levene's Test for Equality of Variances'. As the significance level (sig.) is at .766, which is greater than .05, the two groups have equal variances. Therefore, you cannot say that there are significant differences between the two groups. Consequently, you do not have any reason to reject the null hypothesis (H$_0$), which states that there are no differences between the population means.

When reporting these results, you should write that there was no significant difference in terms of perceived satisfaction by female customers (M = 3.75; SD = 1.619) and male customers (M = 4.32; SD = 1.611); t (80) = -1.508, p = .766.

Table 4.2: Independent Samples Test

		Levene's Test for Equality of Variances		t-test for Equality of Means						
		F	Sig.	t	df	Sig. (2-tailed)	Mean Difference	Std. Error Difference	95% Confidence Interval of the Difference	
									Lower	Upper
satisfaction	Equal variances assumed	.089	.766	-1.508	78	.136	-.571	.379	-1.326	.183
	Equal variances not assumed			-1.510	55.641	.137	-.571	.378	-1.330	.187

One-way ANOVA

The research objective: To investigate whether there was significant difference in the perceived satisfaction towards CTRL between three different groups of customers, with lower, medium and higher income.

Independent variable: 'income_cat' (containing three categories of income).

Dependent variable: The perceived satisfaction, '**satisfaction**'.

Step 1

First, you need to create the income categories. As explained in **Chapter 2**, you can recode 'income' into categorical data. You can create a new variable named '**income_cat**' and recode '**1**' in '**income**' into '**1**' in '**income_cat**', '**2**' in '**income**' into '**2**' in '**income_cat**', '**3**' in '**income**' into '**2**' in '**income_cat**', '**4**' in '**income**' into '**3**' in '**income_cat**' and '**5**' in '**income**' into '**3**' in '**income_cat**'. Remember to define the variable as '**Nominal**' on '**Variable View**'.

Step 2

In the menu '**Analyze**', choose '**Compare means**' and then '**One-way ANOVA**' as shown in **Figure 4.5**.

Figure 4.5: 'Analyze' Menu in SPSS with 'Compare Means' and 'One-Way ANOVA' Submenus

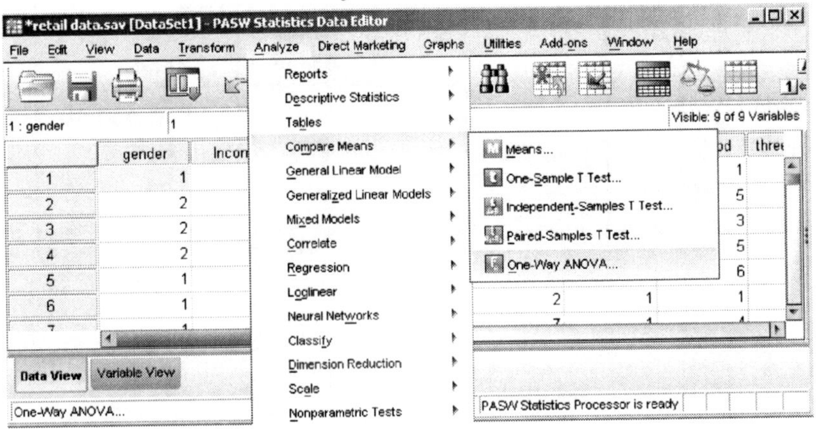

Step 3

In the new window that pops up as shown in **Figure 4.6**, highlight 'satisfaction' and move it into the box '**Dependent List**'.

Figure 4.6: 'One-Way ANOVA' Dialogue Box

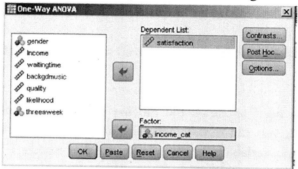

Then, choose the variable '**income_cat**' from the list of variables and move it into the box labelled '**Factor**'. Next, click on the '**Options**' button and then click on '**Descriptive**' and '**Homogeneity of variance test**'. Click on '**Continue**' and '**OK**'.

Tables **4.3, 4.4** and **4.5** show the SPSS output.

Table 4.3: Descriptives

	N	Mean	Std. Deviation	Std. Error	95% Confidence Interval for Mean		Min.	Max.
					Lower Bound	Upper Bound		
Lower	37	3.76	1.535	.252	3.25	4.27	1	7
Medium	31	4.16	1.772	.318	3.51	4.81	1	7
Higher	12	4.00	1.595	.461	2.99	5.01	1	6
Total	80	3.95	1.630	.182	3.59	4.31	1	7

Table 4.4: Test of Homogeneity of Variances

Satisfaction

Levene's Statistic	df1	df2	Sig.
.890	2	77	.415

Table 4.5: ANOVA

Satisfaction

	Sum of Squares	df	Mean Square	F	Sig.
Between Groups	2.796	2	1.398	.520	.597
Within Groups	207.004	77	2.688		
Total	209.800	79			

Interpretation

Table 4.3 presents descriptive statistics for the three groups. It provides information about sample sizes, their respective means, standard deviations, minimum and maximum value etc.

Table 4.4 relates to the assumption about equality of variances. In this case, the significance value of Levene's Statistic is greater than .05 at .415, so you can assume homogeneity of variances between the two groups of consumers.

Finally, in the ANOVA tests as shown in **Table 4.5**, the significance of the F-test is equal to .520 – this is larger than .05 – thus you cannot reject H_0. It means that there is no significant difference in perceived customer satisfaction. Consequently, you need to accept that there are no significant differences between the groups analysed in terms of their perceived satisfaction towards CTRL.

When presenting these results, you should write that there were no significant differences between the analysed groups F = .520, p > .05. Therefore, there is no need to proceed to *post hoc* tests.[7]

---

7   For details on *post hoc* tests, refer to **Chapter 5**.

REFERENCES

Field, A.P. (2000). *Discovering Statistics: Using SPSS for Windows*. London: Sage.

Levine, D.M., Stephan, D., Krehbiel, T.C. & Berenson, M.L. (2001). *Statistics for Managers - Using Microsoft Excel*, 3rd ed. New Jersey: Prentice Hall.

Pallant, J. (2007). *SPSS Survival Manual*, 3rd ed. Maidenhead: Open University Press.

Sirkin, R.M. 2006). *Statistics for the Social Sciences*, 3rd ed. London: Sage.

Stevens, J. (1996). *Applied Multivariate Statistics for the Social Sciences*, 3rd ed. New Jersey: Routledge.

CHAPTER 5
MANOVA

Kieran Flanagan

INTRODUCTION

Multivariate Analysis of Variance (MANOVA) is used to determine the difference between two or more groups with two or more dependent variables. The univariate equivalent, ANOVA, is used where one dependent variable is linked with one, two or more independent groups as discussed in **Chapter 4**. In all situations, the dependent or response variables are required to be continuous and normally distributed.

In the multivariate situation, the dependent variables often are correlated. Research questions that require a difference to be determined across more than one dependent variable, such as sales by volume and sales by value, while being given customer types as an independent variable, are typical. In this case, the independent variables (the customer types) are categorical. Nevertheless, the independent variables also can be scale variables (the shopping value of the previous trip, for example) and will be included in the analysis section.

TYPES OF MANOVA

There are three different types of MANOVA:

◆ **Hotelling's T MANOVA:** This is similar to the t-test used to compare two groups. The difference in Hotelling's T MANOVA is that there are multiple dependent variables while, in a t-test, there is only one dependent variable. *Note:* The number of the independent variables is only ONE and it is dichotomous;

- ◆ **One-way MANOVA:** This is similar to the one-way ANOVA. The difference is that, in one-way MANOVA, there are multiple dependent variables while, in one-way ANOVA, there is only one dependent variable. *Note:* In one-way MANOVA, the number of the independent variables is only ONE and it is multi-level nominal;

- ◆ **Factorial MANOVA:** This is similar to the factorial ANOVA. The difference is that, in factorial MANOVA, there are multiple dependent variables while, in factorial ANOVA, there is only one dependent variable. *Note:* The number of independent variables is multiple and it can be dichotomous or multi-level nominal.

The following section is focused on discussion of the factorial MANOVA.

DESCRIPTION OF FACTORIAL MANOVA

In outlining a statistical test, null and alternative hypotheses need to be identified. For the MANOVA situation based on categorical data, the null hypothesis (H_0) would be that there is no difference between group means for the dependent variables.

This section of the chapter includes examples of each term related to the shopping example mentioned. The total value of the shopping and the number of items purchased on a shopping trip are the dependent variables. These need to be clearly distinguished from the univariate ANOVA, for which the mean of more than two independent variables is compared with one dependent variable. Thus MANOVA would be used in determining whether there is a difference in sales volume and sales value based on the different customer types.

For MANOVA, there are means for each group within the independent variables. One of the independent variables is a grouping or treatment variable, which is the one that you are interested in observing the differences for the dependent variables.

The grouping or treatment variable would be customer type – for example:

♦ Brand-loyal shoppers, who do not change their products;

♦ Relatively-price conscious shoppers, who sometimes change their products;

♦ Barnacle shoppers,[8] who always switch for a bargain.

Knowing the gender of the customer and their frequency of shopping enables a reduction in the grouping variation and a clearer picture of the customer to be created. These independent variables are known as blocking variables or blocks, as they allow the researcher to reduce the variability of the groups.

The null hypothesis in this situation is divided into three groups for the different types of shoppers. Within each of these groups, the gender of the shopper and their frequency of shopping makes up the 24 cells as seen in **Table 5.1** below.

Table 5.1: Subdivision of Customer Types by Frequency and Gender

Shopper Type / Gender	Brand loyal	Relatively price conscious	Barnacle
Male	Once weekly	Once weekly	Once weekly
	Twice weekly	Twice weekly	Twice weekly
	Three times weekly	Three times weekly	Three times weekly
	Four or more times weekly	Four or more times weekly	Four or more times weekly
Female	Once weekly	Once weekly	Once weekly
	Twice weekly	Twice weekly	Twice weekly
	Three times weekly	Three times weekly	Three times weekly
	Four or more times weekly	Four or more times weekly	Four or more times weekly

---

[8]   The term 'barnacle' is used in the sense of Stephen Baker's book, *The Numerati* (2008: 51).

These cells have at least two numerical values: one for each of the dependent variables. For example, one cell consists of brand loyal female shoppers who only shop once a week, while another cell includes male 'barnacle' shoppers who shop three times weekly. Each of these cells has two values for the dependent variables.

The combination of these values has the number of items purchased and the total amount spent as the dependent variables. Once this subdivision is established, a statistical test may be carried out.

There are four widely-used test statistics:

♦ Wilks' Lambda;
♦ Hotelling's Trace;
♦ Pillai's Trace;
♦ Roy's Largest Root.

In general, a test statistic is used to determine differences between dependent or response variables. The four test statistics for MANOVA mentioned above may be transformed or approximated by an F value which is widely calculated and available for determination of critical values to determine the conclusion of the hypothesis. This transformation/approximation is performed in SPSS and the significance of the F value also is calculated, so that the result of the test may be determined. Concerning the F values and significance levels, higher values for the former and smaller values for the latter indicate the likelihood of the alternative hypothesis being true. Among them, the most popular test is Wilks' Lambda.

For the null hypothesis, there are now eight entries for each of the three shopping categories. For example, the combination of the number of items bought and their total value – sales volume and sales value, respectively – would be the same for barnacle males who are brand-loyal, relatively price-conscious and who only shop once a week.

The alternative hypothesis is that there is a difference in the mean value of sales by volume and sales by value for some of the independent variables.

The use of three independent variables – gender, shopping frequency and shopper type – enables an experimental framework to be created and analysed. Within each of these three categorical variables, some variation is expected to occur among the number of items bought and the total value of the shopping. The idea behind this statistical test is to determine whether the variation within each category is due to chance or if it is actually a difference based on the independent variables. In practical terms, some variation in the number of items purchased is expected between two female groups who shop once weekly and are brand loyal. The test should determine when one category is different from another and the outcome is not due to the natural variation mentioned.

REASONS FOR USING MANOVA

When the data is intrinsically multivariate, one MANOVA test is preferred to many univariate ANOVAs. If the data has two or more variables that can be classified as response or dependent variables, MANOVA should be used. Given a single ANOVA, the likelihood of a Type I error – where the sample data suggests that the means are different when the population data states they are not – is specified in advance of the test. This is known as the significance level. Where the multivariate response variables are correlated, performing a series of ANOVAs often magnifies the Type I probability, so that the researcher has no control over the outcome. Using one MANOVA solves this problem.

In the earlier sales example, the performance of two separate ANOVAs for the sales by volume and sales by value may be problematic as these variables are likely to be correlated. Using one MANOVA ensures that the independent variables and dependent variables are included in one analysis and one significance level set for the complete test, rather than having two separate values for individual ANOVAs.

Table 5.2: Comparison of States of Nature with Sample Data

Sample Result	State of Nature	
	Population means same	Population means different
Sample means same	Correct result	Type II error
Sample means different	Type I error	Correct result

In choosing a sample, care must be taken to ensure that randomness is possible. The significance level is normally taken as 5% or .05. The probability of a Type II error can be calculated only after the test has been conducted and is called the power of the test. If the sample chosen is not random, the outcome may not be as significant as is claimed. The results from a sample may be inferred only to the population when the sample is random.

The interaction effects may be included in a MANOVA model. Given three independent variables, a model may not produce conclusive results for the independent variables taken separately, although combinations of them may produce some effect. In the shopping example, the sales volume and sales value were provided for the gender, shopping frequency and type of shopper. The interaction effects would examine, for example, whether there is a difference between male and female shoppers of the three types. This provides useful information for the researcher in many situations. An example is the article by Meyer-Waarden & Benavent (2009) that explores purchasing behaviour of consumers before and after joining a loyalty programme.

DATA REQUIREMENTS

In order for the results of the MANOVA test to be valid, a reasonable sample size should be used. The actual number depends on the particular circumstances but, in general, at least 20 values should be used in each of the cells mentioned above (Hair *et al.*, 2010). More details on sample size are available in Hair *et al.* (2010: 453). Thus, in the sales example above, at least 480 people would be needed so that statistically significant results could be obtained with 20 people in each of the 24 cells. In general, larger samples are more likely to represent the population, so the researcher always should attempt to obtain a large sample to ensure their results are valid.

For small samples of less than 20 in each cell, the multivariate normal distribution is required, which can be examined by using Mahalanobis distance (see **Chapter 3**). If a large sample size is used, this may not be an essential requirement.

The relationships between dependent variables need to be linear, which means that, when two variables are plotted against each other on a scattergraph, the result is approximately a straight line. In the table below, which relates to the worked example in the next section, **'satisfaction'** and **'likelihood'** are positively correlated and the significance level of .000 shows that it is extremely unlikely that the relationship is not linear.

Table 5.3: Correlation between 'satisfaction' and 'likelihood'

		satisfaction	likelihood
satisfaction	Pearson Correlation	1	.626[**]
	Sig. (2-tailed)		.000
	N	80	80
likelihood	Pearson Correlation	.626[**]	1
	Sig. (2-tailed)	.000	
	N	80	80

** Correlation is significant at the 0.01 level (2-tailed).

If a small sample is used, the possibility of **outliers** needs to be addressed. An outlier is an extreme value that may have undue influence on the result of means and variances. Plotting scattergraphs of all independent variables against dependent variables usually identifies this problem. However, in the multivariate situation, care must be taken that interaction effects may be hidden within the data – scattergraphs do not point these out. The Mahalanobis distance is useful in this context (Pallant, 2007: 278).

The dependent variables need to be moderately correlated. If low correlation exists between the dependent variables, the researcher would be better placed performing a series of ANOVAs rather than one MANOVA. If very high correlation exists between the dependent variables – above 0.8, for example – then the analysis may not be valid. This is referred to as multicollinearity and some examination of the dependent variables, such as correlations, should indicate those variables that may be included.

A WORKED EXAMPLE FOR FACTORIAL MANOVA

Research objective: To investigate the differences of perceived satisfaction and likelihood to shop in CTRL in the next week between male and female customers of different income.

    Dependent variables: 'satisfaction' and 'likelihood'.

    Categorical variables: 'gender' and 'income_cat'.

Step 1

Click on '**A**nalyse', then on '**G**eneral Linear Model' and then on '**M**ultivariate', as shown in **Figure 5.1**.

Figure 5.1: 'Analyze' Menu in SPSS with 'General Linear Model' and 'Multivariate' Submenus

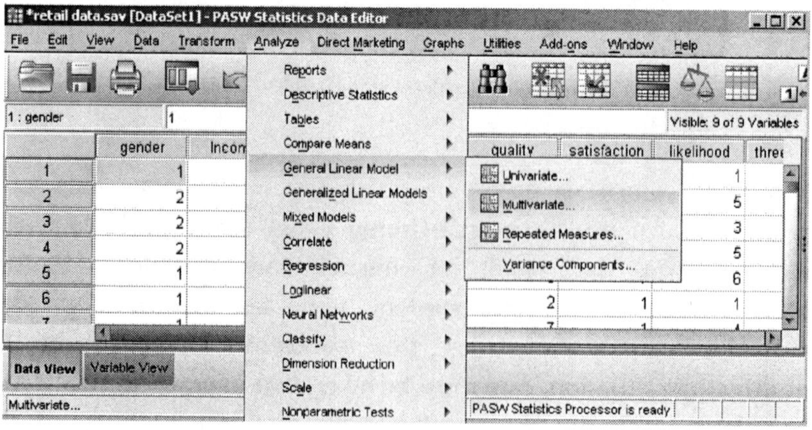

Step 2

A dialogue box appears as in **Figure 5.2**. Move **'satisfaction'** and **'likelihood'** from the list of variables into the box **'Dependent Variables'**, and **'gender'** and **'income_cat'** into the box **'Fixed Factor(s)'**.

Figure 5.2: 'Multivariate' Dialogue Box

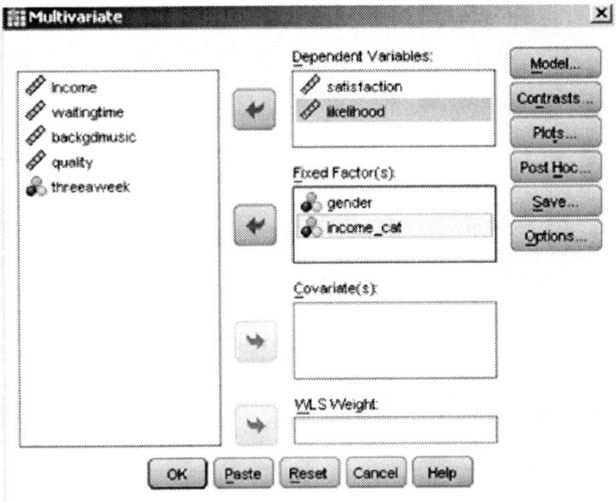

Step 3

In **'Model'**, make sure that **'Full factorial'** is chosen. The **'Contrasts'** option is used to modify these characteristics but normally is not needed.

Step 4

Profile plots may be created by clicking on **'Plots'**. Once the following window pops up as in **Figure 5.3**, move **'income_cat'** into the box under **'Horizontal Axis'** and **'gender'** into the box under **'Separate Lines'**.

Figure 5.3: 'Multivariate: Profile Plots' Dialogue Box

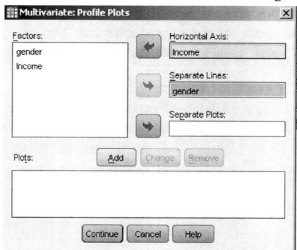

Step 5

Ignore the '**Post Hoc**' option at this stage.

Step 6

'**Option**' is used to determine the differences that have been identified in the main model. In the example, as shown in **Figure 5.4**, choose '**income_cat**' from '**Factor and Factor interactions**' and move it into the box under '**Display Means for**'. Choose '**Descriptive statistics**', '**Estimates of effect size**', and '**Homogeneity tests**'.

Figure 5.4: 'Multivariate: Options' Dialogue Box

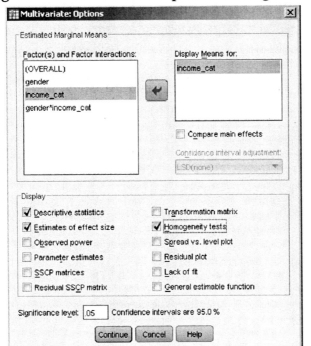

Step 7

The '**Save**' option is used to add the output data from the model so that statistical assumptions may be tested. For advanced users, this would be standard but these requirements are beyond the scope of this book. By selecting '**OK**', MANOVA will be performed.

Step 8

If, in the output, significant F values are detected, then conduct *post hoc* tests by moving the independent variables into the box '**Post Hoc Tests for**' and tick at least '**Bonferroni**' and '**Tukey**', as shown in **Figure 5.5**.

Figure 5.5: 'Multivariate: Post Hoc Multiple Comparisons for Observed Means' Dialogue Box

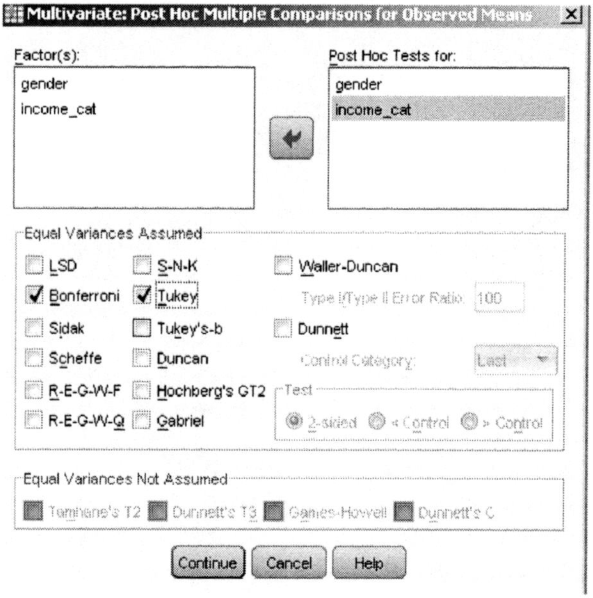

The output data from the SPSS Viewer for the MANOVA is composed of five parts:

♦ Number of values in each category for the independent variables;

♦ Output from the multivariate tests;

♦ Output for Levene's tests;

♦ Output from tests for individual dependent variables;

♦ Output of the post-hoc tests.

Each of these will be described for the worked example.

Table 5.4: Independent Variables Included in the MANOVA Analysis

Between-Subjects Factors

		Value Label	N
gender	1	female	52
	2	male	28
income_cat	1	lower	37
	2	medium	31
	3	higher	12

Table 5.4 gives the number of values in each of the independent variables. There are 52 females and 28 males while the lower, medium and higher income categories have 37, 31 and 12 members respectively. Earlier, it was mentioned that a requirement of the analysis is that there should be at least 20 values in each cell in order to protect from outliers and to ensure that the results of the analysis are valid. Combinations of gender and income category would be the basis of these cells. Taking the extreme example of the higher income category, it can be seen that both cells of male and female high earners will not meet this criterion. Thus further analysis is required in order to test the validity of the output below.

Table 5.5 shows the overall effect of each independent variable and the interaction components on the combination of dependent variables. This is the result of the MANOVA analysis.

Table 5.5: Multivariate Test Output[c]

Effect		Value	F	Hypothesis df	Error df	Sig.	Partial Eta Squared
Intercept	Pillai's Trace	.856	217.825[a]	2.000	73.000	.000	.856
	Wilks' Lambda	.144	217.825[a]	2.000	73.000	.000	.856
	Hotelling's Trace	5.968	217.825[a]	2.000	73.000	.000	.856
	Roy's Largest Root	5.968	217.825[a]	2.000	73.000	.000	.856
gender	Pillai's Trace	.059	2.285[a]	2.000	73.000	.109	.059
	Wilks' Lambda	.941	2.285[a]	2.000	73.000	.109	.059
	Hotelling's Trace	.063	2.285[a]	2.000	73.000	.109	.059
	Roy's Largest Root	.063	2.285[a]	2.000	73.000	.109	.059
income_cat	Pillai's Trace	.107	2.082	4.000	148.000	.086	.053
	Wilks' Lambda	.895	2.090[a]	4.000	146.000	.085	.054
	Hotelling's Trace	.117	2.097	4.000	144.000	.084	.055
	Roy's Largest Root	.104	3.852[b]	2.000	74.000	.026	.094
gender * income_cat	Pillai's Trace	.127	2.510	4.000	148.000	.044	.064
	Wilks' Lambda	.877	2.480[a]	4.000	146.000	.046	.064
	Hotelling's Trace	.136	2.451	4.000	144.000	.049	.064
	Roy's Largest Root	.084	3.099[b]	2.000	74.000	.051	.077

a. Exact statistic. b. The statistic is an upper bound on F that yields a lower bound on the significance level. c. Design: Intercept + gender + income_cat + gender * income_cat.

The multivariate tests examine each factor's effect on the dependent groups. In **Table 5.5**, both '**gender**' and '**income_cat**' are shown not to be significant on the basis of these calculations. The significance value of .109 for the variable of '**gender**' (on the basis of the Wilks' Lambda) indicates that equality for the dependent variables on the basis of the sample is taken as being true. Thus you can confidently state that, on the sample evidence, no difference is found between perceived satisfaction levels of female and male customers. This is the case because a value above the critical significance level of .05 was found for this independent variable.

The significance value for '**income_cat**' is .085, which is greater than .05. Therefore, it can be concluded that there is no significant difference of perceived satisfaction between customers of different income levels. However, it is interesting to find that there is a significant effect of the interaction of '**gender**' and '**income_cat**' – the Wilks' Lambda is at a significance level of .046.

In MANOVA, each dependent variable is assumed to have similar variances for all groups (all 'cells' as discussed earlier). Levene's Test tests such an assumption and it is robust in the presence of departures from normality. Since both significance levels are greater than .05 (see **Table 5.6**), you can conclude that the groups have homogeneity of variances. The assumption is important for MANOVA, which is not the case for ANOVA.

Table 5.6: Levene's Test of Equality of Error Variances[a]

	F	df1	df2	Sig.
likelihood	1.004	5	74	.421
satisfaction	.217	5	74	.954

Tests the null hypothesis that the error variance of the dependent variable is equal across groups.

a. Design: Intercept + gender + income_cat + gender * income_cat

Table 5.7: Tests of Between-Subject Effects

Source	Dependent Variable	Type III Sum of Squares	df	Mean Square	F	Sig.	Partial Eta Squared
Corrected Model	likelihood	34.315[a]	5	6.863	2.703	.027	.154
	satisfaction	24.258[b]	5	4.852	1.935	.099	.116
Intercept	likelihood	887.097	1	887.097	349.413	.000	.825
	satisfaction	928.454	1	928.454	370.297	.000	.833
gender	likelihood	1.031	1	1.031	.406	.526	.005
	satisfaction	3.569	1	3.569	1.423	.237	.019
income_cat	likelihood	19.352	2	9.676	3.811	.027	.093
	satisfaction	7.289	2	3.644	1.453	.240	.038
gender * income_cat	likelihood	12.430	2	6.215	2.448	.093	.062
	satisfaction	15.532	2	7.766	3.097	.051	.077
Error	likelihood	187.873	74	2.539			
	satisfaction	185.542	74	2.507			
Total	likelihood	1385.000	80				
	satisfaction	1458.000	80				
Corrected Total	likelihood	222.188	79				
	satisfaction	209.800	79				

a. R Squared = .154 (Adjusted R Squared = .097)

b. R Squared = .116 (Adjusted R Squared = .056)

When significant results are found for the multivariate tests in **Table 5.5**, it is common to seek the effect of the individual independent variables on the dependent variables, as shown in **Table 5.7**. In the example, the two independent variables 'gender' and 'income_cat', and the interaction of '**gender**' and '**income_cat**' are not significant on the dependent variable '**satisfaction**', which indicates there is no difference of customer satisfaction between the groups. However, 'income_cat' has a significance level less than .05, indicating there is a significant difference of likelihood to shop at CTRL between people with different income levels.

One more piece of information provided in the table is the Eta-squared, which is the proportion of the total variability of the dependent variables accounted for by the variation of the independent variable. In this case for example, gender accounts for about 0.5% of the variability in the '**likelihood**' variable and 1.9% in the '**satisfaction**' variable.

POST HOC TESTS

If the significance of the F test is less than .05, it suggests that there is a significant effect of the independent variable(s) on the dependent variable(s). Then it is recommended to proceed with some *post hoc* tests to determine which group means differ significantly from others. Pair-wise multiple comparison tests for example are popular, to help specify the exact nature of the overall effect determined by the F test. In this example, **Table 5.8** shows there is significant difference in consumers' likelihood to shop at CTRL between two groups of consumers with higher and lower levels of income. The significance level for the Bonferroni test is at .046, while the *post hoc* tests correspond with the previous tests.

Table 5.8: Multiple Comparisons

Dependent Variable		(I) income_cat	(J) income_cat	Mean Difference (I-J)	Std. Error	Sig.	95% Confidence Interval	
							Lower Bound	Upper Bound
satisfaction	Tukey HSD	lower	medium	-.40	.386	.548	-1.33	.52
			higher	-.24	.526	.889	-1.50	1.01
		medium	lower	.40	.386	.548	-.52	1.33
			higher	.16	.538	.952	-1.13	1.45
		higher	lower	.24	.526	.889	-1.01	1.50
			medium	-.16	.538	.952	-1.45	1.13
	Bonferroni	lower	medium	-.40	.386	.892	-1.35	.54
			higher	-.24	.526	1.000	-1.53	1.05
		medium	lower	.40	.386	.892	-.54	1.35
			higher	.16	.538	1.000	-1.16	1.48
		higher	lower	.24	.526	1.000	-1.05	1.53
			medium	-.16	.538	1.000	-1.48	1.16

Dependent Variable		(I) income_cat	(J) income_cat	Mean Difference (I-J)	Std. Error	Sig.	95% Confidence Interval	
							Lower Bound	Upper Bound
likelihood	Tukey HSD	lower	medium	-.89	.388	.062	-1.82	.04
			higher	-1.31*	.529	.040	-2.58	-.05
		medium	lower	.89	.388	.062	-.04	1.82
			higher	-.42	.542	.717	-1.72	.87
		higher	lower	1.31*	.529	.040	.05	2.58
			medium	.42	.542	.717	-.87	1.72
	Bonferroni	lower	medium	-.89	.388	.073	-1.84	.06
			higher	-1.31*	.529	.046	-2.61	-.02
		medium	lower	.89	.388	.073	-.06	1.84
			higher	-.42	.542	1.000	-1.75	.90
		higher	lower	1.31*	.529	.046	.02	2.61
			medium	.42	.542	1.000	-.90	1.75

Based on observed means. The error term is Mean Square(Error) = 2.539.

* The mean difference is significant at the .05 level.

CONCLUSION

An explanation of MANOVA (Multivariate Analysis of Variance), particularly factorial MANOVA, was provided in this chapter, showing the links with ANOVA. A worked example was used to provide a description of the technique. A number of reasons for using MANOVA, as distinct from ANOVA, also were included. The analysis of sample data was performed in this chapter using SPSS to illustrate the technique and the output was explained, including *post hoc* tests.

REFERENCES

Baker, S. (2008). *The Numerati: How They'll Get My Number and Yours*. London: Jonathan Cape.

Hair, J.F., Black, W.C., Babin, B.J. & Anderson, R.E. (2010). *Multivariate Data Analysis*, 7th ed. Upper Saddle River, NJ: Prentice-Hall.

Meyer-Waarden, L. & Benavent, C. (2009). 'Grocery Retail Loyalty Program Effects: Self Selection or Purchase Behaviour Change?', *Journal of the Academy of Marketing Science*, 37: 345-58.

Pallant, J. (2007). *SPSS Survival Manual*, 3rd ed. Maidenhead: Open University Press.

CHAPTER 6
CORRELATION AND REGRESSION

Dónal O'Brien & Pamela Sharkey Scott

INTRODUCTION

A correlation is a measure of the linear relationship between two variables. It is used when a researcher wishes to describe the strength and direction of the relationship between two normally continuous variables. The statistic obtained is Pearson's product-moment correlation (r), and SPSS also provides the statistical significance of r. In addition, if you need to explore the relationship between two variables while statistically controlling for a third, partial correlation can be used. This is useful when it is suspected that the relationship between two variables may be influenced, or confounded, by the impact of a third variable.

Correlations are a very useful research tool but they do not address the predictive power of variables. This task is left to regression. Regression is based on the idea that you first must have some valid reasons for believing that there is a causal relationship between two or more variables. A well-known example is the consumer demand for products and the consumers' level of income. If consumers' income increases, then demand for normal goods such as cars and foreign travel will increase.

In regression analysis, a predictive model needs to fit to both the data and the model. Then you can use the result to predict values of the dependent variable (DV) from one or more independent variables (IVs). In straightforward terms, simple regression seeks to predict an outcome from a single predictor, whereas multiple regression seeks to predict an outcome from several predictors.

CORRELATION

Correlation analysis is useful when you are attempting to establish whether a relationship exists between two variables. For example, you might be interested in whether there is a relationship between the IQ level of students and their academic performance. It might be expected that high levels of IQ are linked to high levels of academic performance. If true, such a relationship indicates a high degree of positive correlation between the IQ level of students and their academic performance.

It is possible also to find correlation where the direction is inverse or negative. For example, a company making computer chips records a high level of defective products and decides to invest in improving its machines and processes. Over time, it sees that, as the investment increased, the number of defective products declined. Therefore, there is a negative correlation between investment and defective products.

It is important to note that correlation provides evidence that there is a relationship between two variables. However, it does not indicate that one variable causes the other. In other words, the correlation between variables A and B could be a result of A causing B, or B causing A, or there could be a third variable that causes both A and B. For this reason, you must always take into account the possibility of a third variable impacting on the observed variables. By using partial correlation, it is possible to statistically control for these additional variables, which allows the possibility of a clearer, less contaminated, indication of the relationship between the two variables of interest.

It is important also to understand the difference between a statistically significant correlation coefficient between the variables and what is of practical significance for the sample. When using large samples, even quite small correlation coefficients can reach statistical significance. For example, although it is statistically significant, the practical significance of a correlation of 0.09 is quite limited. In a case like this, you should focus on the actual size of Pearson's r and the amount of shared variance between the two variables. To interpret the strength of the correlation coefficient, it is advisable to take into account other research that has been conducted in that particular area.

SPSS can calculate two types of correlation. First, it gives a simple bivariate correlation between two variables, also known as zero order correlation. Second, SPSS can explore the relationship between two variables, while controlling for another variable – this is called partial correlation. In this book, only zero order correlation will be discussed.

The two most popular correlations are Spearman's and Pearson's product-moment correlation coefficients. The difference between them is that Pearson's product-moment correlation deals with interval or ratio data, while Spearman rank-order can deal with ordinal data apart from interval and ratio data.

REGRESSION

Regression is particularly useful in understanding the predictive power of the independent variables on the dependent variable, once a causal relationship has been confirmed. To be precise, regression helps you understand to what extent the change of the value of the dependent variable causes the change in the value of the independent variables, while other independent variables are held unchanged.

Simple and Multiple Regressions

In simple linear regression, the outcome or dependent variable Y is predicted by only one independent or predictive variable. Their relationship can be expressed in a math equation as follows:

$$Y = \alpha + \beta X + e \quad \textbf{(6.1)}$$

where: Y is the dependent variable;
α is a constant amount;
β is the coefficient;
X is the independent variable;
e is the error or the 'noise' term that reflect other variables to have an effect on Y.

It should be stressed that, in very rare cases, the dependent variable can be explained only by one independent variable. Thus, to avoid omitted variable bias, multiple regression is applied. Its math equation is as follows:

$$Y = \alpha + \beta_1 X_1 + \beta_2 X_2 + \cdots \beta_n X_n + e \quad (6.2)$$

where Y, α and e remain as the dependent variable, a constant amount and the error respectively. But clearly, the number of the independent variables is now more than one.

Multiple regression is not just a technique on its own. It is, in fact, a family of techniques that can be used to explore the relationship between one continuous dependent variable and a number of independent variables or predictors. Although multiple regression is based on correlation, it enables a more sophisticated exploration of the interrelationships among variables. This makes it suitable for investigating real-life, rather than laboratory-based, research questions.

An important point must be made here. There is a temptation to see regression as a shortcut through the forest of quantitative analysis. Just stick the variables into regression, wait for the answers, and off you go. Job done! This, unfortunately, is not the case. There must be a sound theoretical and conceptual reason for the analysis and, in particular, for the order of the variables entering the equation. Therefore, it is vital to spend time on the research process prior to undertaking regression analysis.

Imagine that the college in which you are enrolled is attempting to analyse the reasons behind students choosing a particular course. The administration team already knows that advertising accounts for 33% of the variation in student enrolment, but a much larger 67% remains unexplained. The college could add a new predictor to the model in an attempt to explain some of the unexplained variation in student numbers. The team decides to measure whether prospective students attended college open days in the year prior to enrolment. The existing model can be extended to include this new variable, and conclusions drawn from the model now are based on two predictors. This analysis will be fulfilled by multiple regression.

Having said that, approached correctly, multiple regression has the potential to address a variety of research questions. It can tell you how well a set of variables is able to predict a particular outcome. For example, if you are interested in exploring how well a set of organisational variables are able to predict organisation performance, multiple regression provides information about the model as a whole and the relative contribution of the variables that make up the model. It also will allow you to measure whether including an additional variable makes a difference, and to control for other variables when exploring the predictive ability of the model.

Some of the main types of research questions that multiple regression can be used to address include:

♦ How well a set of variables can predict a particular outcome;

♦ Identification of the best predictor of an outcome amongst a set of variables;

♦ Whether a particular variable is still able to predict an outcome when the effects of another variable are controlled for.

Assumptions behind Multiple Regression

Multiple regression makes a number of assumptions about the data, and is important that these are met. These assumptions are:

♦ Sample size;

♦ Multicollinearity of independent variables;

♦ Linearity;

♦ Absence of outliers;

♦ Homoscedasticity;

♦ Normality.

Tests of these assumptions are numerous so this chapter will only look at a few of the more important ones.

Sample size

There are a number of recommendations for a suitable sample size for multiple regression analysis (Tabachnick & Fidell, 2007). As a simple rule, calculate the following two values:

$$104 + m$$
$$50 + 8m$$

where m is the number of independent variables, and take whichever is the largest as the minimum number of cases required.

For example, four independent variables would require at least 108 cases, calculated as the greater of $[104 + 4 = 108]$ or $[50 + 8*4 = 82]$, while eight independent variables would require at least 114 cases, being the greater of $[104 + 8 = 112]$ or $[50 + 8*8 = 114]$.

Stepwise regression (see below) needs at least 40 cases for every independent variable (Pallant, 2007).

However, when any of the following assumptions is violated, larger samples are required.

Multicollinearity of Independent Variables

Any two independent variables with a Pearson correlation coefficient greater than 0.9 between them will cause problems. Remove independent variables with a tolerance value less than 0.1. A tolerance value is calculated as $1 - R_i^2$, which is reported in SPSS.

Linearity

Standard multiple regression only looks at linear relationships. You can check this roughly using bivariate scatterplots of the dependent variable and each of the independent variables.[9]

Absence of outliers

Outliers, or extreme cases, can have a very strong effect on a regression equation. They can be spotted on scatterplots in the early stages of analysis.

There are also a number of more advanced techniques for identifying problematic points. These are very important in multiple

---

[9]   More advanced methods include examining residuals.

regression analysis, where you are interested not only in extreme values but in unusual combinations of independent values.

Homoscedasticity

This assumption is similar to the assumption of homogeneity of variance with ANOVAs.

It requires that there be equality of variance in the independent variables for each value of the dependent variable. You can check this in a crude way with the scatterplots for each independent variable against the dependent variable. If there is equality of variance, then the points of the scatterplot should form an evenly-balanced cylinder around the regression line.

Normality

The residuals[10] of the dependent variables should be normally distributed. The independent variables do not have the normality assumption. Refer to **Chapter 3** on how to test univariate and multivariate normality.

Types of Multiple Regression

There are three types of multiple regression:
- **Standard:** All of the independent (or predictor) variables are entered into the equation simultaneously;
- **Hierarchical:** The independent variables are entered into the equation in the order specified by the researcher based on the researcher's theoretical approach;
- **Stepwise:** The researcher provides SPSS with a list of independent variables and then allows the program to select which variables it will use and in which order they go into the equation, based on statistical criteria. This approach has received a lot of criticism and is not covered in any detail in this chapter.

---

[10]   Residuals are the differences between the obtained value and the predicted value of the dependent variable in the regression model.

A WORKED EXAMPLE

Research Question: What is the predictive power of '**backgdmusic**', '**quality**' and '**waitingtime**' on customer's perceived satisfaction towards CTRL, a clothing store?

The Steps for Standard Multiple Regression in SPSS

Step 1

From the menu at the top of the screen, click on '**Analyze**', then '**Regression**', then '**Linear**'.

Figure 6.1: 'Analyze' Menu in SPSS with 'Regression' and 'Linear' Submenus

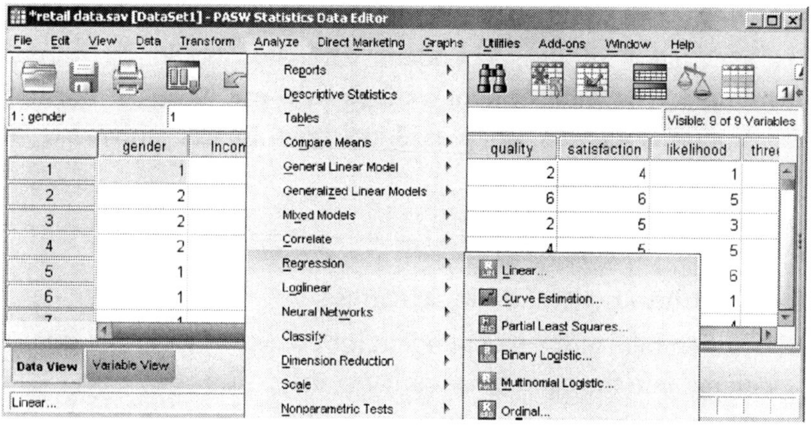

Step 2

Once you are in the window as shown in **Figure 6.2**, click on the continuous dependent variable, '**satisfaction**' and move it into the box '**Dependent**'. Move the variables you want to control into the box '**Independent(s)**' under Block 1 of 1. The variables you normally want to control are demographic variables such as age, gender, income. In this case, move '**gender**' and '**income**' into the box '**Independent(s)**'.

Figure 6.2: 'Linear Regression' Dialogue Box (1)

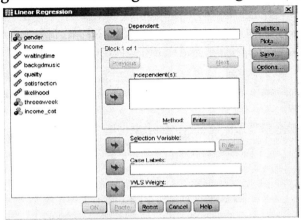

Step 3

Then click on the button marked '**Next**'. This will give a second independent variables box to enter the second block of variables under '**Block 2 of 2**.[11] Move the independent variables, '**waitingtime**', '**backgdmusic**' and '**quality**' into the box '**Independent(s)**'. For '**Method**', make sure '**Enter**' is selected. This will give you standard multiple regression.

Figure 6.3: 'Linear Regression' Dialogue Box (2)

Linear Regression

Dependent: satisfaction

Block 2 of 2

Independent(s):
waitingtime
backgdmusic
quality

Method: Enter

Selection Variable:

Case Labels:

WLS Weight:

gender
income
waitingtime
backgdmusic
quality
likelihood
threeaweek
income_cat

Statistics...
Plots...
Save...
Options...

---

[11] If you do not want to control the variables of '**gender**' and '**income**', you can move all independent variables, in this case, '**gender**', '**income**', '**waitingtime**', **backgdmusic**', and '**quality**' into the box '**Independent(s)**'.

Step 4

Click on the '**Statistics**' button. Then tick the boxes marked '**Estimates**', '**Confidence Intervals**', '**Model fit**', '**Descriptives**', '**Part and partial correlations**' and '**Collinearity diagnostics**'.

In the '**Residuals**' section, tick the '**Casewise diagnostics**' and '**Outliers outside 3 standard deviations**'. Click on '**Continue**'.

Figure 6.4: 'Linear Regression: Statistics' Dialogue Box

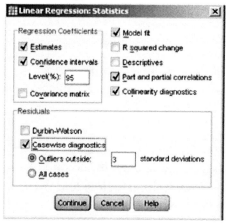

Step 5

If there is missing data, click on the '**Options**' button. In the '**Missing Values**' section, click on '**Exclude cases pairwise**' as shown in **Figure 6.5**.

Figure 6.5: 'Linear Regression: Options' Dialogue Box

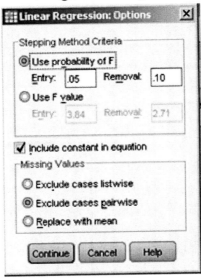

Step 6

Click on the '**Save**' button.

In the next section, labelled '**Distances**', tick the '**Mahalanobis**' box and '**Cook's**' (see **Figure 6.6**). Click on '**Continue**' and then '**OK**'.

Figure 6.6: 'Linear Regression: Save' Dialogue Box

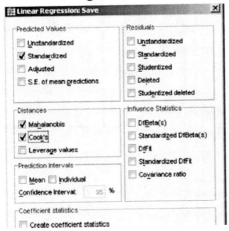

A selected view of the output follows.

Table 6.1: Model Summary[c]

Model		R	R^2	Adjusted R^2	Std. Error of the Estimate	Change Statistics				
						R^2 Change	F Change	df1	df2	Sig. F Change
dimension0	1	.224[a]	.050	.026	1.609	.050	2.037	2	77	.137
	2	.484[b]	.234	.183	1.473	.184	5.930	3	74	.001

a. Predictors: (Constant), gender. Income. b. Predictors: (Constant), gender, income, waitingtime, backgdmusic, quality. c. Dependent Variable: satisfaction.

Table 6.2: ANOVA[c]

Model		Sum of Squares	df	Mean Square	F	Sig.
1	Regression	10.541	2	5.271	2.037	.137[a]
	Residual	199.259	77	2.588		
	Total	209.800	79			
2	Regression	49.161	5	9.832	4.529	.001[b]
	Residual	160.639	74	2.171		
	Total	209.800	79			

a. Predictors: (Constant), Income, gender.
b. Predictors: (Constant), Income, gender, quality, waitingtime, backgdmusic.
c. Dependent Variable: satisfaction.

Table 6.3: Coefficients[a]

Model		Unstandardised Coefficients		Standardised Coefficients	t	Sig.	95.0% Confidence Interval for B		Correlations			Collinearity Statistics	
		B	Std. Error	Beta			Lower Bound	Upper Bound	Zero-order	Partial	Part	Tolerance	VIF
1	(Constant)	2.792	.613		4.556	.000	1.572	4.012					
	gender	.565	.377	.166	1.498	.138	-.186	1.316	.168	.168	.166	1.000	1.000
	Income	.193	.145	.148	1.333	.186	-.095	.481	.150	.150	.148	1.000	1.000
2	(Constant)	.048	.861		.056	.956	-1.667	1.763					
	gender	.463	.352	.136	1.316	.192	-.238	1.164	.168	.151	.134	.963	1.038
	Income	.106	.145	.082	.731	.467	-.183	.395	.150	.085	.074	.831	1.203
	waitingtime	.277	.108	.264	2.555	.013	.061	.493	.287	.285	.260	.967	1.034
	backgdmusic	.175	.143	.139	1.223	.225	-.110	.461	.150	.141	.124	.804	1.245
	quality	.332	.105	.332	3.150	.002	.122	.542	.308	.344	.320	.931	1.074

a. Dependent Variable: satisfaction.

Interpretation of the Output from Hierarchical Multiple Regression

In **Table 6.1**, two models are listed:

♦ **Model 1** refers to the control variables that were entered in Block 1 of 1 (**'gender'** and **'income'**);

♦ **Model 2** includes all the variables that were entered in both blocks (**'gender'**, **'income'**, **'waitingtime'**, **'backgdmusic'**, and **'quality'**).

Evaluate the Model

First, check the R^2 values in the third column. After the variables in Block 1 have been entered, the overall model explains 5% of the variance (.050 x 100%). After Block 2 variables also have been entered, the model in its entirety explains 23.4% (.234 x 100%). Do not forget that the R^2 value in Model 2 includes all of the five variables, not just the three from the second block.

Now you want to establish how much of this overall variance is explained by the variables of interest after the effects of gender and income are removed. To do this, you must look at the column labelled **'R^2 change'**. In **Table 6.1**, on the line marked **'Model 2'**, the R square change value is .184, which means that **'waitingtime'**, **'backgdmusic'**, and **'quality'** explain an additional 18.4% of **'satisfaction'**, even when statistically controlling for the effects of gender and income. This is a statistically significant contribution, as indicated by the Sig. F Change value for this line (.001).

Evaluate Each of the Independent Variables

To find out how well each of the variables predicts the dependent variable, look at the row **'Model 2'** in **Table 6.3**, which contains a summary of the results, with all the variables entered into the equation. In the **'Standardise Coefficient Beta (β)'** column, two variables make a statistically significant contribution. The t of each coefficient β needs to be greater than 2 or less than -2; and the sig. level less than .05.

In the example, **'waitingtime'** has a β of .264 at a sig. level of .013, and t=2.555 and **'quality'** has a β of .332 at a sig. level of .002, and t=3.150. Therefore, you can conclude that, the more customers are pleased with the waiting time at the check-out and the quality of the merchandise, the higher satisfaction is perceived by customers. Neither **'gender'**, **'income'** and **'backgdmusic'** can predict customers' perceived satisfaction significantly.

REFERENCES

Pallant, J. (2007). *SPSS Survival Manual*. Maidenhead: Open University Press.
Tabachnick, B.G. & Fidell, L.S. (2007). *Using Multivariate Statistics*, 5th ed. Boston: Pearson.

CHAPTER 7
LOGISTIC REGRESSION

Siobhan Mc Carthy & Helen Xiaohong Chen

INTRODUCTION

Many decisions involve a question of whether to do something, such as a company outsourcing its call centre or a consumer choosing a service. Such decisions can be modelled with a dummy variable equal to 1 if the activity is done ('**Yes**' is coded as '**1**') or with a dummy variable equal to 0 if it is not done ('**No**' is coded as '**0**') (Schmidt, 2005). This variable is referred to as a dummy dependent variable – a dichotomous dependent variable, to be precise – and thus it needs to be treated differently (Brooks, 2008: 511).

The question arises as to what models are used when the regression model you are estimating has a dichotomous dependent variable – to be exact, a **Yes/No** variable. There are various regression models[12] that could be used, but this chapter's focus is on the logistic regression model.

Students ask why ordinary least squares (OLS) regression cannot be used for dichotomous dependent variable models and why it is necessary to use the logistic model at all. The use of OLS for dichotomous dependent variable models results in various problems, including non-normality, heteroscedasticity and non-independence (Gujarati, 1995; Schmidt, 2005). However, the problem is much more basic than this: least squares analysis does not make sense in the context of a dichotomous dependent variable, which can

---

[12] Including the linear probability model (LPM), the Probit model, the Tobit model and the double hurdle model (see Gujarati, 2003 and Carroll, Mc Carthy & Newman, 2005 for further details).

only take the values 1 or 0. It is impossible to graph a line to this data or talk about how close a line is to such data (Schmidt, 2005: 364).

DESCRIPTION OF THE TECHNIQUE

Logistic regression is similar to OLS regression, but logistic regression models are superior to OLS for limited dependent variable models. A different estimation technique known as maximum likelihood estimation is used for logistic regression.[13]

A logistic regression does not predict the dichotomous variable itself but the probability of the result '1' (**Yes,** or doing something) *versus* '0' (**No,** or not doing something) and thus estimates the expected likelihood of an event occurring: in precise terms, the log odds ratio (Cameron, 2005).

It can be expressed in a mathematical equation as:

$$Logistic\ [P(x)] = log\ \frac{P(x)}{1-P(x)}\ (7.1)$$

where

P(x) is the probability of event x occurring;

1- P(x) is the probability of event x not occurring.

For example, the logistic regression might estimate whether a consumer is purchasing the latest version of a mobile phone. **Equation 7.1** states that a logistic regression model is based on the calculation of the log odds ratio – that is, estimating the log of the ratio of the probability of the consumer purchasing the mobile phone divided by the probability of the consumer not purchasing the mobile phone. Logistic regression analysis then can estimate the coefficients for the independent variables. This can be expressed in a mathematical equation as:

$$Logistic\ [P(x)] = log\frac{P(x)}{1-P(x)} = \alpha + \beta_1 x_1 + \beta_2 x_2 + \ldots \beta_n x_n\ (7.2)$$

Where: α = constant;

---

[13]   The method of maximum likelihood estimates the parameter values that maximise the probability or likelihood of observing the outcomes actually obtained (Hill *et al.,* 2012; Gujarati, 1995).

x_n = the nth independent variable;

β_n = is the nth independent variable coefficient.

A critical question is how to interpret the independent variable coefficients β_n. Only in two aspects is the interpretation of the independent variable coefficients the same as OLS: (1) the sign of the coefficient – that is, whether the coefficient has a positive or negative sign; (2) the significance of the independent variable coefficients as measured by the t statistic.[14] However, it is crucial NOT to interpret the magnitude of the independent variable coefficients in the same way as coefficients estimated using OLS. The magnitude of the independent variable coefficients represents the strength of relationship between the dichotomous dependent variable and the independent variables. Technically, the coefficients of the logistic model are the effects of a one unit change in the independent variable on the log odds ratio (Cameron, 2005). The interpretation of this is far from straightforward. Therefore, the so-called 'marginal effects' must be calculated.

Returning to **Equations 7.1** and **7.2**, the marginal effects can be interpreted as the percentage change in the probability event of $P(X)$ occurring, given a one unit change in any of the independent variables.[15] The conversion of the independent variable coefficients to marginal effects is quite a complex mathematical procedure but is widely available in econometric software.

---

[14] Since the method of maximum likelihood is used, the standard normal z statistic is used instead of the t-statistic as, when the sample size is large, the t-statistic converges to the normal distribution. In addition, pseudo-R^2 is the measure of goodness of fit used instead R^2 and the likelihood ratio (LR) is used instead of the F test statistic for maximum likelihood estimation (Gujarati, 2003; Dougherty, 2002).

[15] If any of the independent variables is itself a dummy variable, the interpretation of the marginal effect is the percentage change in the probability event $P(X)$ occurring, given the difference in percentage points between the two groups within the independent variable dummy (Cameron, 2005).

HOW TO USE LOGISTIC REGRESSION

The use of dichotomous dependent variable models has grown in the last 25 years and now is extremely common in economics modelling (Cameron, 2005). However, dichotomous dependent variable models and, in particular, logistic regression are now also popular for marketing research. Logistic regression is used in predicting the probability of Yes/No questions: for example, consumers buying a product (or not), or consumers joining a loyalty scheme (or not).

In order to illustrate how the logistic regression model can be used, it is useful to take the previous example of whether a consumer purchases the latest version of mobile phone or not.[16] This could be expressed using the mathematical equation below as:

$$Logistic \ [P(x)] = \alpha + \beta_1 Age + \beta_2 Income + \beta_3 Gender + \beta_4 OwnsMobile \quad (7.3)$$

where: α = constant;

Age = age of the consumer;

Income = income level of the consumer;

Gender = dummy variable: whether the consumer is male or female;

OwnsMobile = whether the consumer currently own a mobile phone;

β_n = is the nth independent variable coefficient.

Using **Equation 7.3** as a typical example of logistic regression, the marginal effect for β_2, for example, would represent the percentage change in the probability of purchasing the new version of the mobile phone, given a one unit change in income of the consumer, while the marginal effect for β_2 would represent the percentage change in the probability of purchasing the new version of the mobile phone between males and females.

---

[16] This assumes a very simplistic form. It is assumed that probability of purchasing the new version of the mobile phone depends on four independent variables: namely, the consumer's age, income level, gender and whether the consumer currently owns a mobile phone.

However, it is also crucial that students review some applications of the logistic models in current marketing and economics literature before attempting a logistic regression. A brief summary of two such applications follows.[17]

The first such application of the logistic model is taken from O'Sullivan (2007: 19-28). The dichotomous dependent variable in this application of the logistic model is whether top management regularly considers certain measures.[18] The independent variable of interest in this model is the firm's performance. One of the main findings is 'the regularity with which a firm's top management considers competitive market measures' is positively associated with the firm's performance. Firms rated as performing 'better than' their competitors are significantly more likely to consider competitive market measures than firms rated as performing 'poorer than' their competitors (O'Sullivan, 2007: 31).

The second application of the logistic model is taken from Borooah & Mangan (2008: 351–370). The application in question is the final model in this paper: namely, how the probability of a person in the labour force being unemployed is influenced by his/her personal characteristics and circumstances. In this logistic regression model, 'the dependent variable $Y_i = 1$ if person i was unemployed and $Y_i = 0$ if person i was employed (employee or self-employed)' (Borooah & Mangan, 2008: 262). Some of the findings, illustrated by the marginal effects, are that the probability of being unemployed falls the higher the level of the person's education qualifications, and also with being female, across all regions of the UK.

DATA REQUIREMENTS

A common question students ask is how much data should be collected to use the logistic regression model. Sample sizes for

---

[17] See *References* to review these applications in full. In addition, see Woodbridge (2009); Brooks (2008); Murray (2006); Studenmund (2006); Cameron (2005) and Pindyck & Rubinfeld (1991) for more examples.

[18] These measures are Financial, Competitive Market, Consumer Behaviour, Consumer Attitudes, Direct customer and Innovativeness.

logistic regressions should be substantially larger than for OLS regressions; ideally, the sample size would be 500 or more for economics research (Studenmund, 2006). The reason for this is that maximum likelihood estimation is asymptotically normal and consistent and asymptotically efficient – that is, it has a normal distribution, is unbiased and has minimum variance for larger sample sizes and the standard errors estimated are asymptotic (Gujarati, 2003; Studenmund, 2006). Nevertheless, the requirement for marketing research seems to be lower; according to Agresti (2007: 138), a minimum of 10 observations per independent variable has been recommended.[19]

The size of the sample is not the only critical requirement to run a logistic regression; there also must be representative data from each binary choice. Using the example in **Equation 7.3**, if the data chosen meant that 99% of the sample consumers did not purchase the new version of the mobile phone, the data does not represent both binary choices.

However, it is important to note that logistic regression suffers from the same data issues as OLS or other types of regression, including missing data, non-random data collection, omitted variables and choice of data source. It is beyond the scope of this chapter to review these issues in detail, so the reader may find Franses (2002) useful in this regard.

A WORKED EXAMPLE

The following worked example focuses on the extent that the probability of a consumer shopping in CTRL, a clothing store, is influenced by factors such as the consumer's gender, the consumer's income level, the quality of the merchandise, the waiting time at the check-out and the background music played in the store. In this logistic regression model, the dependent variable $Yi = 1$ if consumer i has shopped at the clothes store previously and $Yi = 0$ if consumer i has not shopped at the clothes store before. It was created by using one dependent variable 'likelihood'. The answer '1' for 'Yes' was

---

[19]   For criticism of this rule of thumb, read Vittinghoff & McCulloch (2006).

coded when respondents selected '**4**' or '**5**' and '**0**' was coded when respondents selected '**1**' or '**2**'.

Research objective: To understand to what extent the quality of the merchandise, the waiting time at the check-out, and the background music predict whether consumers will shop in CTRL in the next week.

Procedure for Data Analysis

The following steps should be followed to estimate a logistic regression in SPSS.

Step 1

Click on '**A**nalyze', and then '**R**egression', and then '**Binary Logistic**'.

Figure 7.1: 'Analyze' Menu in SPSS with 'Regression' and 'Binary Logistic' Submenus

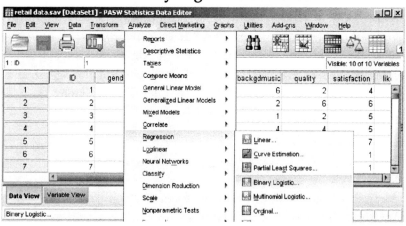

Step 2

In the window that pops up next, as shown in **Figure 7.2**, choose the binary variable '**yes_no**' (which was coded as '**1**' for 'Yes, shopping in CTRL in the next week' *versus* '**0**' for 'No, not shopping in CTRL in the next week') and move it into the '**D**ependent' box.

Figure 7.2: 'Logistic Regression' Dialogue Box (1)

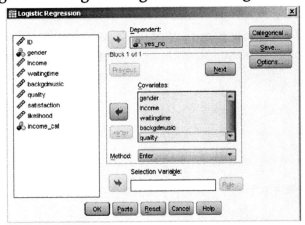

Step 3

Choose the independent variables **'gender'**, **'income'** **'waitingtime'**, **'backgdmusic'**, and **'quality'** and move them into the box **'Covariates'**.[20] Make sure to choose **'Enter'**.

Figure 7.3: 'Logistic Regression' Dialogue Box (2)

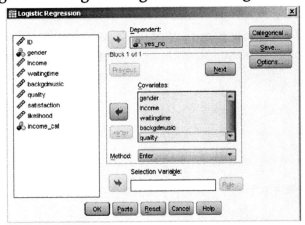

---

[20]   By doing this, **'gender'** and **'income'** are not control variables. If you want to control them, you should move only **'gender'** and **'income'** into **'Covariates'** in Block 1 of 1. Click on 'Next' and then move **'waitingtime'**, **'quality'** and **'backgdmusic'** into **'Covariates'** in Block 2 of 2.

Step 4

As **'gender'** is the categorical variable, click on **'Categorical'** and then move **'gender'** into the **'Categorical covariates'** box, as shown in **Figure 7.4.** Click on **'Continue'**.

Figure 7.4: 'Logistic Regression: Define Categorical Variables' Dialogue Box

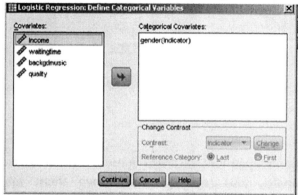

Step 5

Click on the **'Option'** button, then select **'Classification plots'**, **'Hosmer Lemeshow goodness of fit'**, **'Casewise listing of residuals'** and **'CI for exp(B)'**, and then click on **'Continue'**. Then click on **'OK'**.

Figure 7.5: 'Logistic Regression: Options' Dialogue Box

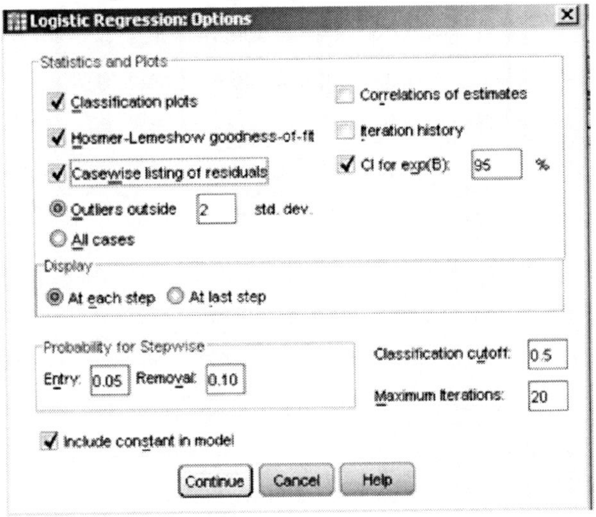

DATA ANALYSIS RESULTS

Presentation of the Output

SPSS presents a lot of outputs based on the requirements set by the researcher. It is recommended that students make their own tables to present only useful findings based on the output file. Nevertheless, in this chapter, the main objective is to help students interpret the tables in the output file. Therefore, the original tables from SPSS are shown in **Tables 7.1, 7.2, 7.3** and **7.4**.

Table 7.1: Model Summary

Step	-2 Log likelihood	Cox & Snell R^2	Nagelkerke R^2
1	92.296[a]	.206	.274

a. Estimation terminated at iteration number 4 because parameter estimates changed by less than .001.

Table 7.2: Omnibus Tests of Model Coefficients

		Chi-square	df	Sig.
	Step	18.408	8	.018
Step 1	Block	18.408	8	.018
	Model	18.408	8	.018

Table 7.3: Hosmer and Lemeshow Test

Step	Chi-square	df	Sig.
1	6.723	8	.567

Interpretation of the Tables

Table 7.1 highlights that the Cox and Snell R^2 is at .206 and the Nagelkerke R^2 is at .274. Together, these measures suggest that between 20.6% and 27.4% of the variability of the dependent variable is explained by the set of independent variables. Note that Nagelkerke R^2 is always larger than Cox and Snell R^2 as it is a different scale.

Omnibus Tests of Model Coefficients in **Table 7.2** is at the significance level of .018, which is less than .05. As Omnibus Tests for Model Coefficients provide an overall indication of how well the model performs over and above the results obtained for Block 0[21] (Pallant, 2007), the model is better than the original guess by SPSS shown in Block 0.

The Chi-square value is 18.408 with 8 degrees of freedom. In the Hosmer and Lemeshow Test in **Table 7.3**, the Chi-square is at 6.723 at the significance level of .567, higher than the significance level of .05, indicating the model is valid.

---

[21] In Block 0, there is no independent variable that is entered in the model. Only the intercept is evaluated in the model.

Table 7.4: Variables in the Equation

	B	S.E.	Wald	df	Sig.	Exp(B)	95% C.I.for EXP(B) Lower	95% C.I.for EXP(B) Upper
gender(1)	-.987	.551	3.210	1	.073	.373	.127	1.097
waitingtime	.150	.177	.716	1	.397	1.162	.821	1.643
backgdmusic	.417	.242	2.970	1	.085	1.518	.944	2.439
quality	.413	.197	4.393	1	.036	1.512	1.027	2.225
Income			3.700	4	.448			
Income(1)	.653	1.047	.389	1	.533	1.921	.247	14.954
Income(2)	-.207	1.102	.035	1	.851	.813	.094	7.051
Income(3)	1.405	1.265	1.235	1	.266	4.077	.342	48.623
Income(4)	.763	1.446	.279	1	.598	2.145	.126	36.494
Constant	-3.287	1.742	3.559	1	.059	.037		

a. Variable(s) entered on step 1: gender, Income, waitingtime, backgdmusic, quality.

According to **Table 7.4**, only 'quality' has significant effect on the possibility that a consumer will shop in CTRL in the next week, as the Wald value is at a significance level of less than .05 and the β for 'quality' is .413, which indicates that consumers who perceive the clothing offered to be of a higher quality are more likely to shop in CTRL in the next week. The Odds Ratio [Exp(β)] for 'quality' is 1.512, indicating the odds of shopping being preferred by customers in the next week is 1.512 times higher when the clothes are perceived to be of higher quality. Other variables such as 'backgdmusic', 'income' and 'waitingtime' have no significant predictive power in the model.[22]

CONCLUSION

Logistic regression is a popular technique used in business research. It does not predict the 0/1 variable itself but the probability of obtaining 1 *versus* 0. The first part of this chapter outlines the logistic regression technique, the uses of logistic regression and its data requirements. The second part of the chapter reviews an application of logistic regression, namely establishing which factors influence whether consumers will shop at a clothes store in the next week.

REFERENCES

Agresti, A. (2007). 'Building and Applying Logistic Regression Models', in Agresti, A. (2007), *An Introduction to Categorical Data Analysis*. Hoboken, NJ: Wiley-Interscience.

Borooah, V.K. & Mangan, J. (2008). 'Education, Occupational Class, and Unemployment in the Regions of the United Kingdom', *Education Economics*, 16(4): 351-70.

Brooks, C. (2008). *Introductory Econometrics for Finance*. Cambridge: Cambridge University Press.

Cameron, S. (2005). *Econometrics*. New York: McGraw-Hill.

---

[22] Unfortunately, SPSS does not offer calculation of marginal effects. If it is required, you can use other software such as LIMDEP for this purpose.

Carroll, J, Mc Carthy, S. & Newman, C. (2005). 'An Econometric Analysis of Charitable Donations in the Republic of Ireland', *The Economic and Social Review*, 36(3): 229-49.

Dougherty, C. (2002). *Introduction to Econometrics*, 2nd ed. Oxford: Oxford University Press.

Franses, P.H. (2002). *A Concise Introduction to Econometrics: An Intuitive Guide*. Cambridge: Cambridge University Press.

Guajarati, D.N. (1995). *Basic Econometrics*, 3rd ed. New York: McGraw-Hill.

Guajarati, D.N. (2003). *Basic Econometrics*, 4th ed. New York: McGraw-Hill.

Hill, R.C., Griffiths, W. & Guay, L. (2012). *Principles of Econometrics*, 4th ed. Asia, Wiley.

Murray, M.P. (2006). *Econometrics: A Modern Introduction*. Boston: Pearson.

O'Sullivan, D. (2007). 'The Measurement of Marketing Performance in Irish Firms', *Irish Marketing Review*, 19(1): 19-28.

Pallant, J. (2007). *SPSS Survival Manual*, 3rd ed. Maidenhead: Open University Press.

Pindyck, R.S. & Rubinfeld, D.L. (1991). *Econometrics Models & Economic Models*, 3rd ed. Singapore: McGraw-Hill.

Schmidt, S.J. (2005). *Econometrics*. New York: McGraw-Hill.

Studenmund, A.H. (2006). *Using Econometrics: A Practical Guide*, 5th ed. Boston: Pearson.

Vittinghoff, E. & McCulloch, C. (2006). 'Relaxing the Rule of Ten Events per Variable in Logistic and Cox Regression', *American Journal of Epidemiology*, 165 (6): 715-718.

Woodbridge, J.M. (2009). *Introductory Econometrics: A Modern Approach*, 4th ed. Scarborough, ON: South Western Cengage Learning.

CHAPTER 8
EXPLORATORY FACTOR ANALYSIS

Daire Hooper

INTRODUCTION

Factor analysis examines the inter-correlations that exist between a large number of items (questionnaire responses) and in doing so reduces the items into smaller groups, known as factors. These factors contain correlated variables and typically are quite similar in terms of content or meaning. Unlike other methods discussed in this book, exploratory factor analysis (EFA) does not discriminate between variables based on whether they are independent or dependent, but rather it is an interdependence technique that does not specify formal hypotheses. It is in this sense that it is 'exploratory' in nature, as it allows the researcher to determine the underlying dimensions or factors that exist in a set of data.

The technique is particularly useful for managerial or academic research in reducing items into discrete dimensions that can be summed or aggregated and subsequently used as input for further multivariate analysis such as multiple regression. It is used extensively also in scale development research to condense a large item pool into a more succinct, reliable and conceptually sound measurement instrument.

Factor analytic techniques typically can be classified as either exploratory or confirmatory; the former of these is addressed within this chapter, using a research example to demonstrate its use.

WHEN WOULD YOU USE FACTOR ANALYSIS?

There are a number of reasons why you would use factor analysis. The first is when you want to determine whether a series of dimensions or factors exist in the data and whether they are interpretable in a theoretical sense. For instance, if you collected data from respondents to determine how committed they were to maintaining employment in their organisation, you might use Allen & Meyer's (1990) 24-item organisational commitment scale. They propose that three sub-dimensions exist within the organisational commitment construct: affective, continuance and normative commitment. Factor analysis can be used to determine whether this three-factor structure is replicable in the dataset – in other words, to ascertain whether employees conceptually classify organisational commitment along these three dimensions.

EFA would examine the inter-correlations between all variables on Allen & Meyer's (1990) scale and, from that, reduce the data into a smaller number of dimensions (factors). The dimensions produced by factor analysis then can be used as input for further analysis such as multiple regression. In the case of the organisational commitment example, each of the items on a dimension could be summed to create an aggregate item and subsequently regressed on a dependent variable such as turnover.

The second reason for using factor analysis would be to refine the number of items on a scale for the purposes of scale development (DeVellis, 2003). Factor analysis allows the researcher to determine the nature and number of latent variables (dimensions/factors) underlying a set of items. One of the critical assumptions associated with scale construction is for items measuring a particular construct to be relatively homogenous or unidimensional (that is, load together on one factor). To meet this end, factor analysis can be used to determine whether one, or multiple, dimensions exist in a set of variables. Scale development is not within the scope of this book; however, interested readers can refer to DeVellis's (2003) or Spector's (1992) comprehensive texts on the subject.

THE DIFFERENCE BETWEEN EXPLORATORY FACTOR ANALYSIS AND PRINCIPAL COMPONENTS ANALYSIS

Too often, principal components analysis (PCA) is referred to as EFA but this is an inaccurate classification. To a novice researcher, both techniques may appear to be the same – particularly with regard to their execution and output in SPSS. However, mathematically and theoretically, they differ considerably.

The widespread adoption of PCA can be attributed to it being the default extraction method in both SPSS and SAS (Costello & Osborne, 2005). Holding this default position has more than likely led to PCA being used mistakenly when EFA is more suitable (Park, Dailey & Lemus, 2002). The goal of PCA is to reduce the measured variables to a smaller set of composite components that capture as much information as possible in as few components as possible. On the other hand, the goal of EFA is to find the underlying structure of the dataset by uncovering common factors. Therefore, EFA accounts for shared variance. This is an important distinction from PCA, as it means EFA is more suitable when exploring underlying theoretical constructs.

There has been much debate over which of these techniques is the true method of factor analysis, with some arguing in favour of EFA (Floyd & Widaman, 1995; Gorsuch, 1990; Snook & Gorsuch, 1989), while others argue there is little difference between the two (Velicer & Jackson, 1990).

Principal axis factoring, a type of EFA, is superior to PCA as it analyses common variance only which is a key requirement for theory development.[23] In addition to this, it is a useful technique for identifying items that do not measure an intended factor or that simultaneously measure multiple factors (Worthington & Whittaker, 2006). For these reasons, exploratory techniques are most important for theory development and will be used here.

---

[23] Theory development is concerned with adding to knowledge and understanding how key constructs or phenomena relate to one another.

DATA REQUIREMENTS

Factor analysis is typically a large sample size technique, with correlations less reliable when small samples are used. Recommendations on appropriate sample sizes for factor analysis vary considerably (Fabrigar et al., 1999). Some have suggested a minimum of 300 cases is required; however, in practice, about 150 should be sufficient (Tabachnick & Fidell, 2007) and as few as 100 cases can be adequate in situations where there are a small number of variables. The items themselves must be interval in nature (for example, Likert scales) and although ideally multivariate normality is a requirement, deviations from this are not usually detrimental to the results. It is also important that the researcher assesses for outliers, as their presence can alter the factor solution (Tabachnick & Fidell, 2007).

WORKED EXAMPLE

Examination of the service quality literature finds that most authors describe service quality as an overall appraisal of a product or service that is dependent on consumers' prior expectations (Grönroos, 1984; Bitner & Hubbert, 1994) and it is this disconfirmation-based definition that prevails most commonly in the literature.

Within the services quality literature, two complementary streams of research have evolved, which can be broadly categorised as being of either the Scandinavian or American tradition. As previously mentioned, both of these schools of thought agree that consumers arrive at an evaluation of service quality that is based upon disconfirmation theory. This being so, prior to consuming a service, consumers hold preconceived ideas of how the service will perform. Once the consumer has experienced the service, they compare performance to their a priori expectations in a subtractive manner to determine their perceptions of service quality.

Parasuraman et al.'s (1985; 1988; 1994) model falls into the American tradition and is the most widely cited service quality model in the literature. Building on the premise that quality perceptions are a function of expectations and performance, they developed the Gaps Model and its associated five-dimension SERVQUAL measurement instrument.

Within the Nordic stream of research, Grönroos (1984) proposed that service quality can be described as a two-factor structure comprising both functional and technical elements. The functional element relates to the *way* in which the service is delivered, while the technical element refers to *what* the consumer receives from the service (Brogowicz *et al.*, 1990). This functional aspect of service delivery has been referred to as peripheral to the process, while the technical element conceptually constitutes the core or outcome components of the service delivery process (Tripp & Drea, 2002). Writings on this model have been mostly theoretical (Ekinci *et al.*, 1998). However, in more recent years, a number of authors have sought to link technical and functional quality dimensions to a variety of constructs such as trust, commitment, satisfaction and loyalty (Lassar *et al.*, 2000; Caceres & Paparoidamis, 2005) and general support has been found for the two dimensional conceptualisation of service quality.

This chapter continues this line of research by examining whether a two-dimensional model of service quality is replicated in a fictitious dataset. The context for this example is a service station and includes 355 cases. Service quality was measured using items developed by Grace and O'Cass (2004), as well as a number of self-developed items. These can be found in **Table 8.1** below. All items were measured using 7-point scales anchored with 'strongly disagree' (1) and 'strongly agree' (7).

Table 8.1 Service Quality Items

The service was delivered promptly

The service here was reliable

The service was efficient

The staff were helpful

The staff were polite

The staff were friendly*

The staff were trustworthy

The service station provided quality service*

The service station provided good service*

The service here suited my needs*

* denotes self-developed items.

Factor Analysis Procedure in SPSS

Having decided these service quality items are to be used, the next stage is actually running the factor analysis. As previously mentioned, a fictitious dataset containing 355 cases was created to demonstrate the technique and can be found on the website **www.oaktreepress.com** (Factor Analysis Dataset.sav).

Step 1

Once you have opened the file in SPSS, select '**A**nalyze', click on '**D**imension Reduction' and then '**F**actor'. At this point, a window will open and you will see all service quality items on the left-hand side (see **Figure 8.1**). Select the variables you wish to include in the analysis and move them over to the '**V**ariables' section on the right-hand side. For this example, move across all 10 items in the dataset.

Figure 8.1: Selecting Variables in Factor Analysis

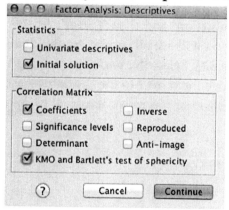

Step 2

Select the '**D**escriptives' button and, in the section marked '**Correlation Matrix**', select '**C**oefficients' and '**K**MO and Bartlett's test of sphericity' and click '**Continue**' (shown in **Figure 8.2**). These are selected to test a number of assumptions associated with factor analysis and will be discussed later.

Figure 8.2: 'Factor Analysis: Descriptives' Dialogue Box

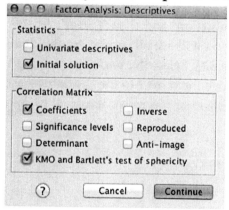

Step 3

Click on the '**Extraction**' button and in the '**Method**' section, make sure '**Principal axis factoring**' is selected from the dropdown box. This is shown in **Figure 8.3** (*Note:* If you were using PCA, you would choose '**Principal components**' here).

In the '**Analyse**' section, make sure '**Correlation matrix**' is selected. Under '**Display**', select '**Unrotated factor solution**' and tick the check box beside '**Scree plot**'. In the '**Extract**' section, there are two options with radio buttons beside each: the first is '**Eigenvalues greater than 1**' as this is the default extraction method in SPSS. For now, leave this as it is and we will return to it later. Click '**Continue**'.

Figure 8.3: 'Factor Analysis: Extraction' Dialogue Box (1)

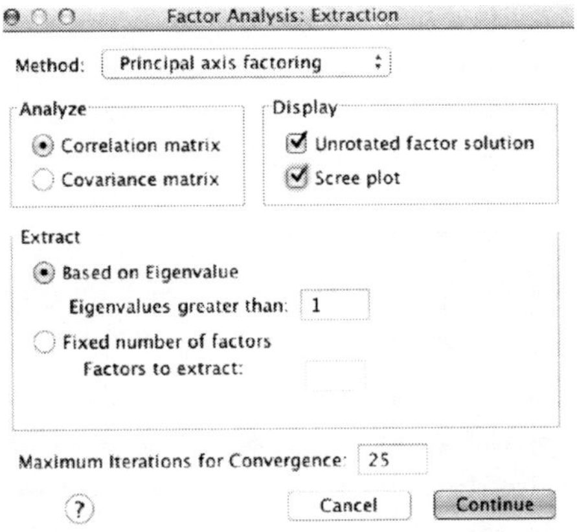

Step 4

Next click on the '**Rotation**' button and select '**Promax**', as shown in **Figure 8.4**. The default is for Kappa 4 here; leave it as it is. You have chosen to use Promax rotation as this is a type of oblique rotation that allows for correlations between factors. There are other oblique rotation methods (for example, Direct Oblimin); however, Promax is generally chosen as it is quicker and simpler.

We believe service quality dimensions will be correlated with one another and this is the rationale for choosing this type of rotation. If you were using principal components, you would choose Varimax rotation as this is an orthogonal rotation technique that maximises the variances of loadings on the new axes.

Figure 8.4: 'Factor Analysis: Rotation' Dialogue Box

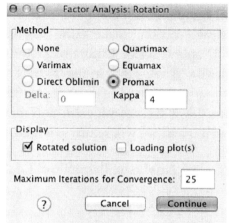

Step 5

As shown in **Figure 8.5**, the next steps are to click on '**Options**' and to make sure the radio button is selected beside '**Exclude cases pairwise**'.

Following this, in the '**Coefficient Display Format**' section, select '**Sorted by size**'. Sorting by size means factor coefficients will be listed from the largest down to the smallest, which will help when interpreting results.

Then select '**Suppress small coefficients**' and enter the value .4 (Hair *et al.*, 2006) in the box beside '**Absolute value below**'. By choosing this option, SPSS will hide coefficients less than .4. This is a useful tool for a number of reasons. First, it helps interpretation, as you can see more clearly where particular items load. Second, it highlights items with loadings less than .4 on all dimensions. When an item does not load on a dimension (that is, it has loadings less than .4 on all dimensions), it may indicate the item is unreliable and, as a result, may be a candidate for deletion. Finally, this also shows whether any items cross-load, meaning that an item is loading on more than one dimension, which would lead us to question the reliability of the item.

Figure 8.5: 'Factor Analysis: Options' Dialogue Box

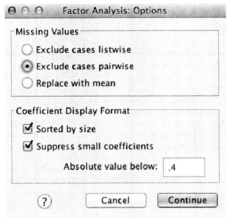

Step 6

Once all of the above have been selected, the next stage is to run the analysis. To do this, click '**Continue**' and '**OK**'. The factor analysis output will open in a second window known as your output file.

Interpretation of Output

Factor analysis produces a considerable amount of output but this should not deter you in its interpretation. In this next section, the important pieces of information will be explained.

Stage 1 – Testing the Assumptions

The first thing you need to do is to look over the Correlation Matrix to ensure you have correlation coefficients greater than .3 in magnitude. If you do not have any correlations over .3, it might indicate factor analysis is not appropriate. In this example, there are quite a number of correlations greater than .3, which tentatively suggests factor analysis is appropriate here (see **Table 8.2**).

Table 8.2: Correlation Matrix

	The service in store was delivered promptly	The service here was reliable	The staff were trustworthy	The staff were friendly	The service here suited my needs	The staff were polite	The service was efficient	The staff were helpful	The service station provided good service	The service station provided quality service
The service in store was delivered promptly	1.000	.656	.445	.515	.339	.582	.674	.465	.451	.345
The service here was reliable	.656	1.000	.502	.671	.419	.573	.573	.473	.520	.405
The staff were trustworthy	.445	.502	1.000	.490	.306	.578	.460	.484	.494	.441
The staff were friendly	.515	.671	.490	1.000	.382	.527	.510	.455	.496	.490
The service here suited my needs	.339	.419	.306	.382	1.000	.454	.401	.333	.528	.400
The staff were polite	.582	.573	.578	.527	.454	1.000	.696	.545	.567	.443

	The service in store was delivered promptly	The service here was reliable	The staff were trustworthy	The staff were friendly	The service here suited my needs	The staff were polite	The service was efficient	The staff were helpful	The service station provided good service	The service station provided quality service
The service was efficient	.674	.573	.460	.510	.401	.696	1.000	.580	.567	.466
The staff were helpful	.465	.473	.484	.455	.333	.545	.580	1.000	.419	.375
The service station provided good service	.451	.520	.494	.496	.528	.567	.567	.419	1.000	.667
The service station provided quality service	.345	.405	.441	.490	.400	.443	.466	.375	.667	1.000

Next, check the value of the Kaiser-Meyer-Olkin Measure of Sampling Adequacy (KMO), which should be either .6 or above. For this example KMO is .904 which is well within acceptable limits (see **Table 8.3** below). The Bartlett's Test of Sphericity should be significant (less than .05) and, in this example, we have met this criterion as the test is significant (p=.000).

Table 8.3: KMO and Bartlett's Test of Sphericity

Kaiser-Meyer-Olkin Measure of Sampling Adequacy		.904
Bartlett's Test of Sphericity	Approx. Chi-Square	1788.071
	df	45
	Sig.	.000

Stage 2 – Deciding on the Number of Factors to Extract

The next decision relates to the number of factors to extract. The number of dimensions selected can be based on a range of criteria and it is widely recommended a variety of approaches are used when making this decision (Fabrigar *et al.*, 1999). According to Tabachnick & Fidell (2007), this stage should take an exploratory approach by experimenting with the different numbers of factors until a satisfactory solution is found. However, in order for you to do this, you will need to familiarise yourself with the different criteria that can be used to determine the number of factors.

The first and most popular method for deciding on the retention of factors is Kaiser's 'eigenvalue greater than 1' criterion (Fabrigar *et al.*, 1999). This rule specifies that all factors greater than '1' are retained for interpretation. This method offers the advantage of being easy to understand and it is also the default method on most programs. Some argue this method oversimplifies the situation and also has a tendency to overestimate the number of factors to retain (Zwick & Velicer, 1986). In fact, this method may lead to arbitrary decisions – for example, it does not make sense to retain a factor with an eigenvalue of 1.01 and then to regard a factor with an eigenvalue of .99 as irrelevant (Ledesma & Pedro, 2007).

A technique that overcomes some of the deficiencies inherent in Kaiser's approach is Cattell's scree test (Cattell & Vogelmann, 1977). The scree test graphically presents the eigenvalues in descending order, linked with a line. This graph is then scrutinised to determine where there is a noticeable change in its shape, which is known as 'the elbow' or point of inflexion. Once you have identified the point at which the last significant break takes place, only factors *above and excluding* this point should be retained.

A priori theory also can drive the process, so if a break was found further along the scree plot and made theoretical sense, then factor analysis could be re-run specifying the appropriate number of factors.

An alternative criterion is to set a predetermined level of cumulative variance and to continue the factoring process until this minimal value is reached. While no absolute threshold has been adopted, for the social sciences Hair *et al.* (2006) state that a minimum of 60% cumulative variance is quite commonly accepted; however, realistically, researchers are happy with 50% to 75% of the variance explained.

A final method is Horn's (1965) parallel analysis. Unfortunately, this method is not built into the SPSS user-interface; however, interested readers can use O'Connor's (2000) syntax if they wish to apply it to their data.

Finally, when deciding upon the number of factors, you are strongly advised against underfactoring (choosing too few factors). This is considered a much more serious error than specifying too many (Cattell, 1978) as it can lead to distortions whereby two common factors are combined into a single common factor thus obfuscating the true factor structure.

Reverting to the example, if you plan to apply Kaiser's 'eigenvalue greater than 1' criterion you would extract only one factor from the dataset. This is determined by examining the Total Variance Explained table (shown below in **Table 8.4**), which shows that the total eigenvalues for the first dimension is 5.469, accounting for 54.70% of the variance extracted, or 49.88% of the shared variance. Looking to the line below this, the second factor has not met the 'eigenvalue greater than 1' criterion as it has an eigenvalue of

.932. As you will recall, Kaiser's 'eigenvalue greater than 1' criterion has been criticised for its relatively arbitrary selection of factors and here is a situation where the second factor possesses an eigenvalue of .932, which is reasonably close to the eigenvalue of 1 cut-off point. Given the closeness of these eigenvalues to 1, you may decide to re-run the analysis specifying a two-dimensional solution. However, for now, proceed by applying each of the other factor extraction criteria to your results as it is recommended to use a combination of criteria to arrive at a final decision.

Table 8.4: Total Variance Explained

Factor	Initial Eigenvalues			Extraction Sums of Squared Loadings		
	Total	% of Variance	Cumulative %	Total	% of Variance	Cumulative %
1	5.469	54.695	54.695	4.988	49.876	49.876
2	.932	9.318	64.013			
3	.693	6.928	70.941			
4	.657	6.565	77.507			
5	.564	5.638	83.145			
6	.507	5.065	88.210			
7	.369	3.686	91.896			
8	.311	3.110	95.007			
9	.263	2.625	97.632			
10	.237	2.368	100.000			

Extraction Method: Principal Axis Factoring.

Next, examine the scree plot (**Figure 8.6**) to find the point of inflexion (elbow). In this example, the most obvious break (point of inflexion) is at Factor 2 (as indicated by the heavy circle in **Figure 8.6**), suggesting a one-dimensional solution is appropriate. However, a second (albeit much smaller) drop in eigenvalues seems to occur between Factor 2 and 3, which may indicate a two-factor solution is appropriate (highlighted by the lighter circle in **Figure 8.6**).

Figure 8.6: Scree Plot for Exploratory Factor Analysis

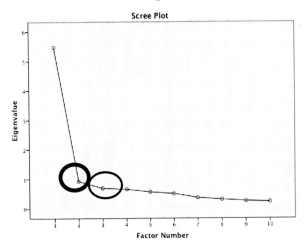

Furthermore, if you apply the cumulative variance criterion, your one factor solution captures 49.88% of the variance, which unfortunately does not meet Hair *et al.*'s (2006) more stringent 60% threshold, or the generally accepted 50% cut-off point. This result, combined with your eigenvalue analysis and scree plot inspection should lead you to consider a two-factor solution.

Coupled with these results, you must bear in mind your *a priori* theoretical framework, which proposed a two factor solution. Therefore, you should re-run the analysis, this time specifying a two-factor solution.

To do this, proceed through all steps described above (Analyse/Dimension Reduction/Factor etc.). However, when you click on 'Extraction', rather than leave the default as 'Based on Eigenvalues greater than 1', click on 'Fixed number of factors' and enter the value '2' in the box beside 'Factors to extract'. Select 'Continue' and 'OK' to re-run the factor analysis.

Figure 8.7: 'Factor Analysis: Extraction' Dialogue Box (2)

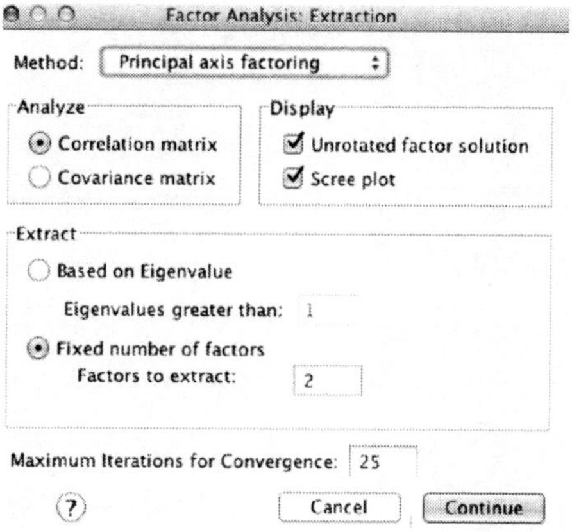

New output will be generated and you will notice the correlation matrix, KMO and Bartlett's Test of Sphericity are all the same as for the original specification. The only difference is that SPSS has produced a two-dimensional solution, rather than a one-dimensional solution. This can be seen by examining the Total Variance Explained table as shown in **Table 8.5**.

Note in the section '**Extraction Sums of Squared Loadings**' that are there two lines of data rather than just one, which reflects the fact that you have constrained the solution to two dimensions. The solution now has accounted for 64.01% of variance, or 55.90% of shared variance in the data. This is a preferable situation to the one-factor solution as when too few factors are included in a model, substantial error is likely (Fabrigar *et al.*, 1999).

Table 8.5: Total Variance Explained (Re-specified Solution)

Factor	Initial Eigenvalues			Extraction Sums of Squared Loadings			Rotation Sums of Squared Loadings
	Total	% of Variance	Cumulative %	Total	% of Variance	Cumulative %	Total
1	5.469	54.695	54.695	5.048	50.476	50.476	4.745
2	.932	9.318	64.013	.541	5.415	55.891	3.986
3	.693	6.928	70.941				
4	.657	6.565	77.507				
5	.564	5.638	83.145				
6	.507	5.065	88.210				
7	.369	3.686	91.896				
8	.311	3.110	95.007				
9	.263	2.625	97.632				
10	.237	2.368	100.000				

Extraction Method: Principal Axis Factoring.

Stage 3 – Factor Rotation and Interpretation

The next stage is to interpret the factors. Principal axis factoring produces slightly different tables to other forms of factor analysis; however, the table you are most interested in is the **Pattern Matrix,** which displays the rotated factor loadings and is used to interpret the dimensions.

However, before beginning interpretation, the first thing you need to check is for cross-loadings. A cross-loading is an item with coefficients greater than .4 on more than one dimension (Hair *et al.,* 2006). To help with this, you requested all loadings less than .4 be suppressed in the output to aid interpretation. The example is free from cross-loadings, as all items load on only one dimension. The second thing you need to check is whether there are items that do not load on any of the factors – that is, they have loadings *less* than .4 on all dimensions. Again, all items load on either the first or the second dimension, providing a nice clean solution to interpret.

If you found items cross-loading or not loading at all, this would suggest they are poor/unreliable items and may need to be deleted from the analysis. If this were to happen, you would need to re-run your analysis without the offending item(s).

Having reached a suitable solution, the next stage is to interpret the factors themselves. This has been referred to by some as a 'black art' as there are no hard or fast rules in naming each dimension. However, there are a number of guidelines that can aid in the process. First, there are two factors and variables load highly on only one factor with no cross-loadings; they are arranged in descending order to help us identify items with substantive loadings. These variables with higher loadings are used to identify the nature of the underlying latent variable represented by each factor.

Table 8.6: Pattern Matrix

	Factor	
	1	2
The service in store was delivered promptly	.902	
The service here was reliable	.760	
The service was efficient	.746	
The staff were polite	.677	
The staff were helpful	.601	
The staff were friendly	.548	
The staff were trustworthy	.464	
The service station provided good service		.870
The service station provided quality service		.791
The service here suited my needs		.443

Extraction Method: Principal Axis Factoring.
Rotation Method: Promax with Kaiser Normalization.

Rotation converged in 3 iterations.

In this example, the variables loading on the first factor all relate to the service process, or the human element of the service delivery. The second dimension contains three items, which appear to be evaluative items whereby the respondents are providing the service with an overall rating.

These two dimensions are in keeping with your proposed theory that stated consumers perceive services along two discrete, yet related, dimensions. The first of these is the functional dimension and corresponds to the way in which the service is delivered. By and large, this is dependent on the service delivery process and frontline employees play a key role here. All items on the first dimension relate to the role of the employee and are in keeping with your understanding of functional service quality and, as such, will be named 'Functional Service Quality'. The items on the second dimension can be regarded as outcome-type items as they refer to 'what' kind of service the customer received and, for this reason, the dimension is named 'Technical Service Quality'.

Reporting Factor Analysis Results

When reporting the results from factor analysis, there are a number of key pieces of information you need to include so a reader can assess the decisions you made. It is essential you report the extraction technique used, the rotation technique used (Promax, Varimax, etc.), the total variance explained, the initial eigenvalues and the rotated eigenvalues. You also need to include a table of loadings showing all values (not just those in excess of .4) in the Pattern Matrix. As an oblique rotation method was used, you should also report the Structure Matrix.

For this example, the results would be described along the following lines:

> Ten service quality items (Grace & O'Cass, 2004) were subjected to principal axis factoring to assess the dimensionality of the data. The Kaiser-Meyer-Olkin was .904, which is well above the recommended threshold of .6 (Kaiser, 1974), and the Bartlett's Test of Sphericity reached statistical significance, indicating the correlations were sufficiently large for exploratory factor analysis.

> Two factors were extracted explaining 64.01% of the variance. This was decided based on eigenvalues, cumulative variance and inspection of the scree plot. Factors were obliquely rotated using Promax rotation and interpretation of the two factors was in keeping with Grönroos's (1984) two-dimensional theory of service quality. Items that load on the first dimension suggests it represents Functional Service Quality and the second dimension suggests it represents Technical Service Quality.

Table 8.7: Pattern Matrix for Coefficients

	Factor	
	1	**2**
	Functional Service Quality	**Technical Service Quality**
The service in store was delivered promptly	**.902**	-.165
The service here was reliable	**.760**	.037
The service was efficient	**.746**	.081
The staff were polite	**.677**	.159
The staff were helpful	**.601**	.076
The staff were friendly	**.548**	.210
The staff were trustworthy	**.464**	.238
The service station provided good service	.011	**.870**
The service station provided quality service	-.045	**.791**
The service here suited my needs	.170	**.443**
% of variance explained	**54.69%**	**9.31%**

Reliability Analysis

If you are to use scales in your research, it is essential that they are reliable. Reliability refers to how free the scale is from random error and is frequently measured using a statistic known as Cronbach's alpha (α).

Cronbach's alpha is a measure of internal consistency, which means the degree to which items in your scale measure the same underlying attribute or construct. Cronbach's alpha ranges from 0 to 1, with higher values indicating high levels of reliability. Nunnally (1978) recommends a minimum of .7; however, alpha values increase with scale length, so checking for unidimensionality *via* exploratory factor analysis is key here.

For this example, you will test the reliability of all items on the Functional Service Quality dimension.

Step 1

To do this, click on '**Analyze**', then '**Scale**' and finally '**Reliability Analysis**'. Move all seven Functional Service Quality variables to the '**Items**' field and click '**Statistics**' (shown in **Figure 8.8** below).

Figure 8.8: 'Reliability Analysis' Dialogue Box

Step 2

In the '**Descriptives for**', section select '**Item**' and '**Scale if item deleted**'. Then click '**Continue**' and then '**OK**'.

Figure 8.9: 'Reliability Analysis: Statistics' Dialogue Box

In the output table named 'Reliability Statistics' (Table 8.8), the first column provides the alpha coefficient, which is .883 and is well above Nunnally's .7 threshold.

Table 8.8: Reliability Statistics

Cronbach's Alpha	N of Items
.883	7

In the table marked 'Item Total Statistics', look to the 'Cronbach's Alpha if Item Deleted' column to determine whether the alpha value would change substantially if you delete particular items. If there are values for some items higher than your Cronbach's alpha, you might want to re-run Cronbach's alpha excluding these items.

For this example, there appear to be no problems here so you can proceed to use this scale in further analysis. In order to do so, you must calculate the total scale scores for each of the dimensions. Summating scales is common practice in research and is done to allow you to perform statistical tests that require continuous variables (correlation, multiple regression, ANOVA and MANOVA all use continuous variables). Before calculating a total score, check that no items on your scale are negatively worded. If items are negatively worded, they will need to be reverse coded in SPSS (that is, if you have a scale ranging from '1' to '7', reverse coding means replacing all '1's with '7', '2's with '6', all the way to '7's replaced with '1'). In this example, all items on the Functional Service Quality scale are positively worded so you can proceed to add all items.

Table 8.9: Item Total Statistics

	Scale Mean if Item Deleted	Scale Variance if Item Deleted	Corrected Item-Total Correlation	Cronbach's Alpha if Item Deleted
The service in store was delivered promptly	37.60	16.643	.697	.863
The service here was reliable	37.67	16.700	.738	.859
The staff were trustworthy	37.78	15.972	.615	.876
The staff were friendly	38.01	15.385	.665	.870
The staff were polite	37.46	17.687	.746	.864
The service was efficient	37.58	16.718	.733	.860
The staff were helpful	37.66	16.297	.624	.873

In SPSS, select '**Transform**' and then '**Compute Variable**'. In '**Target variable**', type in a name for the new summated item you want to create. For this example, enter '**FunctionalSQ**'. From the list on the left-hand side, select the first item from the Functional Service Quality scale (prompt) and move it to the '**Numeric Expression**' box and click on the '**+**' on the calculator. Proceed in this manner until all Functional Service Quality items are in the box (see **Figure 8.10**).

Figure 8.10: Computing Variables

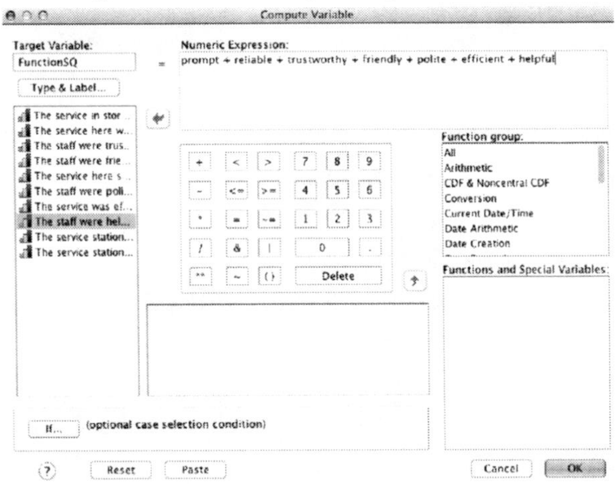

Once all items have been entered, select '**OK**' and SPSS will generate a new item which will be listed after all other variables in the dataset. This new item can now be used in other analyses that require continuous variables.

SUMMARY

This chapter has introduced the reader to exploratory factor analysis and has demonstrated how it can be used to assess the dimensionality of a dataset. In particular, a two-dimensional factor structure of service quality was found.

The following summarises the steps undertaken when using factor analysis:

♦ Ensure your sample size is sufficiently large (minimum of 150 or 10 cases per item). Items should be interval in nature.

♦ Check your correlation matrix to ensure there are a reasonable number of correlations coefficients greater than .3.

♦ Check KMO and Bartlett's Test of Sphericity. KMO should be .06 or above. Bartlett's Test of Sphericity should reach significance at the .05 level or higher.

♦ Choose an extraction method. For this example, Principal Axis Factoring was chosen. If you were purely interested in data reduction, rather than theory building, Principal Components Analysis would be more suitable.

♦ Choose a rotation method. Here, an oblique method (Promax) was chosen as factors were expected to be correlated with one another. Where factors are not expected to correlate, orthogonal methods such as Varimax can be used.

♦ Decide on the number of factors to extract. The default in SPSS is Kaiser's 'eigenvalue greater than 1' criterion. It is recommended that a number of factor extraction methods are used. The example given here relied on *a priori* theory, the scree plot and the percentage of variance extracted.

♦ Using the high loading items, interpret dimensions in the Pattern Matrix.

♦ To use dimensions for further analysis, sum items to form a composite item.

REFERENCES

Allen, N.J. & Meyer, J.P. (1990). 'The Measurement and Antecedents of Affective, Continuance and Normative Commitment to the Organization', *Journal of Occupational Psychology* 63(1): 1-18.

Bitner, M.J. & Hubbert, A.R. (1994). 'Encounter Satisfaction *versus* Overall Satisfaction *versus* Quality', in Rust, R.T. & Oliver, R.L. (eds.), *Service Quality: New Directions in Theory and Practice*, pp. 72-94. Thousand Oaks, CA: Sage.

Brogowicz, A.A., Delene, L.M., & Lyth, D.M. (1990). 'A Synthesised Service Quality Model with Managerial Implications', *International Journal of Service Industry Management* 1(1): 27-45.

Caceres, R.C. & Paparoidamis, N.G. (2005). 'Service Quality, Relationship Satisfaction, Trust, Commitment and Business-to-Business Loyalty', *European Journal of Marketing* 41(7/8): 836-867.

Cattell, R.B. & Vogelmann, S. (1977). 'A Comprehensive Trial of the Scree and KG Criteria for Determining the Number of Factors', *Multivariate Behavioral Research* 12(3): 289.

Cattell, R.B. (1978). *The Scientific Use of Factor Analysis in Behavioral and Life Sciences*. New York: Plenum

Costello, A.B. & Osborne, J.W. (2005). 'Best Practices in Exploratory Factor Analysis: Four Recommendations for Getting the Most From Your Analysis', *Practical Assessment, Research & Evaluation* 10(7): 1-9.

DeVellis, R.F. (2003). *Scale Development: Theory and Applications*, 2nd ed. Thousand Oaks, CA: Sage.

Ekinci, Y., Riley, M. & Fife-Schaw, C. (1998) 'Which School of Thought? The Dimensions of Resort Hotel Quality', *International Journal of Contemporary Hospitality Management* 10(2): 63-67.

Fabrigar, L.R., MacCallum, R.C., Wegener, D.T., & Strahan R. (1999). 'Evaluating the Use of Exploratory Factor Analysis in Psychological Research', *Psychological Methods* 4(3): 272-299.

Floyd, F.J. & Widaman, K.F. (1995). 'Factor Analysis in the Development and Refinement of Clinical Assessment Instruments' *Psychological Assessment* 7(3): 286-299.

Gorsuch, R.L. (1990). 'Common Factor Analysis versus Component Analysis: Some Well and Little Known Facts', *Multivariate Behavioral Research* 25(1): 33-39.

Grace, D. & O'Cass, A. (2004). 'Examining Service Experiences and Post-Consumption Evaluations', *Journal of Services Marketing* 18(6): 450-461.

Grönroos, C. (1984). 'A Service Quality Model and its Marketing Implications', *European Journal of Marketing* 18(4): 36-44.

Hair, J.S., Black, W.C., Babin, B.J., Anderson, R.E., & Tatham, R.L. (2006). *Multivariate Data Analysis*, New Jersey: Prentice-Hall.

Horn, J.L. (1965). 'A Rationale and Test for the Number of Factors in Factor Analysis', *Psychometrika* 30(2): 179-183.

Kaiser, H.F. (1974). 'An Index of Factorial Simplicty, *Psychometrkia* 39: 31-36.

Lassar, W.M., Manolis, C., & Winsor, R.D. (2000). 'Service Quality Perspectives and Satisfaction in Private Banking', *Journal of Services Marketing* 14(3): 244-271.

Ledesma, R.D. & Pedro, V.M. (2007). 'Determining the Number of Factors to Retain in EFA: an easy-to-use computer program for carrying out Parallel Analysis', *Practical Assessment, Research & Evaluation* 12(2): 1-11.

Nunnally, J.C. (1978). *Psychometric Theory*. New York: McGraw-Hill.

O'Connor, B.P. (2000). 'SPSS and SAS Programs for Determining the Number of Components Using Parallel Analysis and Velicer's MAP Test', *Behavior Research Methods, Instruments and Computers* 32(3): 396-402.

Parasuraman, A., Zeithaml, V.A., & Berry, L. (1985). 'A Conceptual Model of Service Quality and Its Implications For Future Research', *Journal of Marketing* 49(4): 441-450.

Parasuraman, A., Zeithaml, V.A., & Berry, L. (1988). 'SERVQUAL: A Multiple Item Scale for Measuring Customer Perceptions of Service Quality', *Journal of Retailing* 64(1): 12-40.

Parasuraman, A., Zeithaml, V.A., & Berry, L. (1994). 'Reassessment of Expectations as a Comparison Standard in Measuring Service Quality: Implications for Further Research', *Journal of Marketing* 58(1): 111-124.

Park, H.D., Dailey, R., & Lemus, D. (2002). 'The Use of Exploratory Factor Analysis and Principal Components Analysis in Communication Research', *Human Communication Research* 28(4): 562-577.

Snook, S.C. & Gorsuch, R.L. (1989). 'Component Analysis Versus Common Factor Analysis: A Monte Carlo Study', *Psychological Bulletin* 106(1): 148-154.

Spector, P.E. (1992). *Summated Rating Scale Construction*. Newbury Park, CA: Sage Publications.

Tabachnick, B.G. & Fidell, L.S. (2007). *Using Multivariate Statistics*. New York: Allyn and Bacon.

Tripp, C. & Drea, J.T. (2002). 'Selecting and Promoting Service Encounter Elements in Passenger Rail Transportation', *Journal of Services Marketing* 16(5): 432-442.

Velicer, W.F. & Jackson, D.N. (1990). 'Component Analysis versus Common Factor Analysis: Some Issues in Selecting an Appropriate Procedure', *Multivariate Behavioral Research* 25(1): 97-114.

Worthington, R.L. & Whittaker, T.A. (2006). 'Scale Development Research: A Content Analysis and Recommendations of Best Practices', *The Counselling Psychologist* 34(6): 806-838.

Zwick, W.R. & Velicer, W.F. (1986). 'Comparison of Five Rules for Determining the Number of Components to Retain', *Psychological Bulletin* 99(3): 432-442.

CHAPTER 9
MULTIDIMENSIONAL SCALING

Donncha Ryan

INTRODUCTION

The technique of multidimensional scaling (MDS) can be applied to a set of objects for which either attribute-based or similarity data is available, thus allowing for the visualisation of these objects in a low dimensional space. It is exploratory in nature.

A researcher may be assessing and representing the perception of similarity in relation to a set of, say, 'n' objects. The perception of similarity between the objects (brands, products, etc.) typically is presented as a matrix of similarities (or distances) that a respondent provides subjectively, or is calculated from available data relating to the objects from which a distance matrix may be computed. Once a distance or similarity matrix is available, the task is to find an appropriate low dimensional (usually 2-D Euclidean) space whereby the objects are represented as points whose configuration reflects the original similarities in terms of their relative inter-point distances on the map. The final map representation is judged in terms of various goodness of fit measures (primarily involving the notion of stress), with a good solution being characterised by a low stress value. A balance between simplicity and parsimony of representation *versus* validity and goodness of fit is sought. This translates to visualisation in two or three dimensions, provided the fit is reasonable.

Depending on the nature of the input data and the transformations applied to it, MDS techniques can be classified broadly as metric or non-metric MDS. Much of their origin was in the field of psychometrics, with some of the earlier approaches developed by Torgerson (1958), Shepard (1962a, 1962b), Kruskal (1964) and others.

In summary, a rather modest data requirement in the form of perceived or derived 'similarity' ratings regarding a set of 'n' objects serves as the input, with the output being a 2-D visualisation of the objects, revealing the similarity structure and allowing for possible interpretation of underlying perception dimensions used by the respondent. MDS thus is useful, particularly when attribute evaluations are difficult to obtain or not available for the objects studied.

DATA REQUIREMENTS

The input data does not suffer from burdensome assumptions such as linearity, normality or other distributional assumptions. The input data type need not be metric and is often only ordinal and subjective in nature. Hypothesis testing is not accommodated by the technique, it being exploratory in nature, emphasising visualisation of objects and the interpretation of underlying dimensions affecting perceived similarity and cognition.

The input data consists generally of a matrix (usually square and symmetric) of similarity (or distances) values concerning the set of objects. Therefore, in the case of 'n' objects, it consists of an 'n by n' square symmetric matrix. The similarity matrix can be examined at the individual respondent or aggregate level.

This chapter will consider a scenario of a researcher exploring the perception of similarity among a set of 'n' objects by an individual respondent.

This might take the form (in this case) of a question such as:

Please rate the following pairs of brands in terms of similarity, using a scale of 1-10; with higher scores indicating greater similarity. Ties are allowed.

Figure 9.1: A Sample Survey Question for Similarity Ratings

Brands	1	2	3	4	5	6	7	8
1								
2								
3								
4								
5								
6								
7								
8								

The extension to several respondents (not pursued in the chapter) may be approached in several ways, including the derivation of individual maps, or developing a joint map or indeed aggregating the similarity matrices and developing an average similarity matrix.

The MDS family of techniques encompasses a wide variety of approaches, ranging from classical scaling, to metric scaling, to non-metric scaling, to Procrustes analysis to multidimensional unfolding, etc. (see, for example, Cox & Cox, 2001). This chapter describes only the approach of non-metric (ordinal) MDS.

WHEN TO USE IT

Possible scenarios include:

♦ You ask the respondent to provide their subjective measures of similarities. Equally, the question may be posed in terms of dissimilarities and the respondent provides distance type evaluations. This involves pair-wise comparisons and ratings and the data collected may be in the form of:

  ◊ **Ratings:** pair-wise ratings – for example, on a scale of 1 to 10, with '1' denoting a least similar pair and '10' being most similar, or in the case of distance data, the opposite applies;

  ◊ **Rankings:** for example, the most similar pair is ranked '1' and the farthest pair is ranked 'r'. Thus, in the case of eight objects, there will be $r = {}^8C_2 = 28$ pairs to consider and the pairs will be ranked from '1' to '28' (ties are possible). This is reversed in the case of distances. Naturally, there is an element of potential respondent fatigue and tedium in being asked to consider all pair-wise comparisons and assigning appropriate rankings;

◊ **Preferences:** here the respondent can rate (for example, on a nine-point scale) or rank the objects in terms of preference. Analysing preference data is possible with multidimensional unfolding algorithms (not pursued in this chapter) using the PREFSCAL algorithm. For an example, see Meulman & Heiser (2005, Chapter 8);

♦ Variable/attribute information is available regarding each object under study. For example, if you have ratings on 10 brands of coffee across variables such as quality, aroma, strength, etc., you can construct a distance matrix between the brands. You then can assess perception and similarity from this distance matrix using metric MDS to derive a 2-D map. Depending on the data type, measures of distance appropriate to the data need to be chosen and there is a wide variety of distance measures available. If the data is measured on different scales, standardisation may be necessary before computing distances. Naturally, if you have such data on a set of objects, you have recourse to using other multivariate techniques such as cluster analysis, principal component analysis, etc.

TYPES OF MDS

MDS can be described broadly as being either metric or non-metric. This chapter focuses on non-metric MDS and presents a worked example, using subjective ratings of similarity rather than attribute-based data.

Assume that you wish to represent a set of, say, 'n' objects as points in a p-dimensional space (where p<n). The data input is a matrix of perceived similarities or proximities. The entries represent the similarity or proximity ratings of the objects or the notional distance between the objects (brands, companies, stimuli, etc.). You

want to obtain a configuration of the objects in a low dimensional (usually 2-D) space. Their positions and their separation on the map should be a reflection of their perceived similarity. These derived distances between the objects in the configuration are to be related in some way to the original distances and, as such, you are faced with matching the two sets of distances: namely, perceived/inputted proximities and map-derived distances. How this matching is achieved technically differentiates the various MDS techniques.

Non-metric MDS

Following Kruskal (1964), you adopt a pragmatic approach as follows: given that the original measures of similarity may be subjective ratings, you demand only that the rank order of the dissimilarity or proximity ratings be preserved in the derived configuration. This basically means that a pair of objects perceived as being most similar by the respondent should be most closely positioned together on the map and so on to the pair perceived as being most dissimilar being most separated on the map. Various algorithms are available, one of the earliest being the Shepard-Kruskal algorithm for non-metric MDS (Cox & Cox, 2001: 69). Basically, you want to find a configuration such that the derived map distances are in the same rank order as the inputted proximity ratings. There is always a means of representing them perfectly in (n-1) dimensions but usually 2-D spaces are used for interpretation purposes.

The task of finding a spatial configuration of the 'n' objects in the p-dimensional space and the resulting inter-object map distances appropriately rank ordered is a minimisation problem.

A variety of measures of the 'goodness of the fit' is available to assess how valid is the resulting map configuration, and they tend to be a variation of a central theme of 'stress'. Typically, a particular MDS algorithm will try to minimise the particular stress formulation adopted by the minimisation algorithm, and report for reference on a variety of other standard stress measures for comparison and assessment of the fit.

A WORKED EXAMPLE

Consider a respondent rating eight radio station brands in terms of their similarity on a scale of 1 to 10.

Research objective: To assess, represent and explore the perceived similarity of the set of brands in an attribute-free approach, with a view to understanding the subjective dimensions underpinning this evaluation. More specifically:

♦ To assess how many dimensions are needed or used;

♦ To identify what those dimensions may be;

♦ To assess the nature of the similarity structure of these eight brands.

This worked example uses MDS as an exploratory device in identifying subjective dimensions in the evaluation of the set of objects without necessary reference to attribute information.

Step 1

Access the data being analysed or indeed type the similarity ratings into the SPSS data editor as shown below in **Figure 9.2**.

Figure 9.2: Data Entry

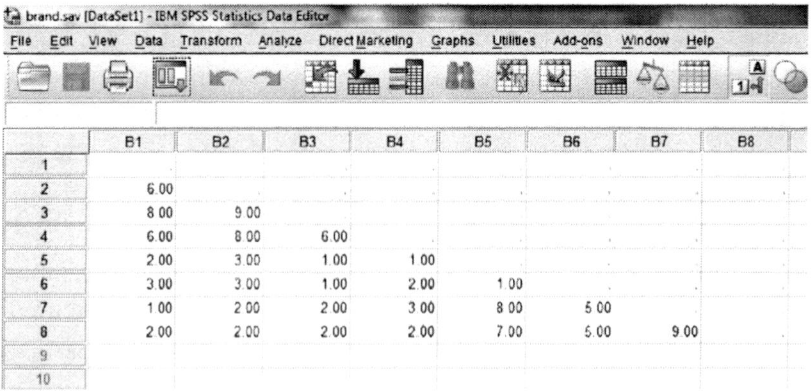

	B1	B2	B3	B4	B5	B6	B7	B8
1								
2	6.00							
3	8.00	9.00						
4	6.00	8.00	6.00					
5	2.00	3.00	1.00	1.00				
6	3.00	3.00	1.00	2.00	1.00			
7	1.00	2.00	2.00	3.00	8.00	5.00		
8	2.00	2.00	2.00	2.00	7.00	5.00	9.00	
9								
10								

Step 2

Click on '**A**nalyze' in the main menu and then on '**Sc**ale' and '**Multidimensional Scaling**', as shown in **Figure 9.3**.

Figure 9.3: 'Main' Menu in SPSS with 'Scale' and 'Multidimensional Scaling' Submenus

| av [DataSet2] - SPSS Data Editor | | | | | | | | |

View Data Transform [Analyze] Graphs Utilities Add-ons Window Help

B1	B2		Reports ▶		B6	B7	B8	var
			Descriptive Statistics ▶					
			Tables ▶					
6 00			Compare Means ▶					
8 00	9 00		General Linear Model ▶					
6 00	8 00		Generalized Linear Models ▶					
2 00	3 00		Mixed Models ▶					
3 00	3 00		Correlate ▶					
1 00	2 00		Regression ▶		5 00			
2 00	2 00		Loglinear ▶		5 00	9 00		
			Classify ▶					
			Data Reduction ▶					
			Scale ▶	Reliability Analysis...				
			Nonparametric Tests ▶	Multidimensional Unfolding...				
			Time Series ▶	Multidimensional Scaling (PROXSCAL)...				
			Survival ▶	Multidimensional Scaling (ALSCAL)...				
			Multiple Response ▶					
			Missing Value Analysis...					
			Complex Samples ▶					

Step 3

The next dialog box in **Figure 9.4** asks about the data format, the number of sources and about the way the data appears in the spreadsheet.

In this case, the data are proximity ratings with one respondent (source) and are stored in lower triangular form as shown above. Proceed by clicking on '**Define**'.

Figure 9.4: 'Multidimensional Scaling: Data Format' Dialogue Box

Step 4

In the next dialogue box, as shown in **Figure 9.5**, enter the variables into the proximities box and proceed to the '**Model**' button.

Figure 9.5: 'Multidimensional Scaling (Proximities in Matrices across Columns)' Dialogue Box

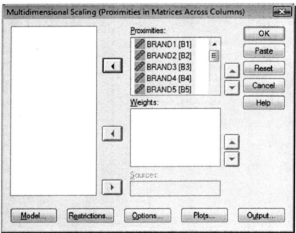

Step 5

In the '**Model**' options indicated in **Figure 9.6**, choose an ordinal MDS with tied proximities (there were several) being allowed to be untied in the configuration. In 'Dimensions', request solutions of two to four dimensions so as to generate a scree plot and to assess the fit for each dimension being considered. The initial data were proximities, so that is indicated by ticking the '**Similarities**' button under '**Proximities**'.

Figure 9.6: 'Multidimensional Scaling: Model' Dialogue Box

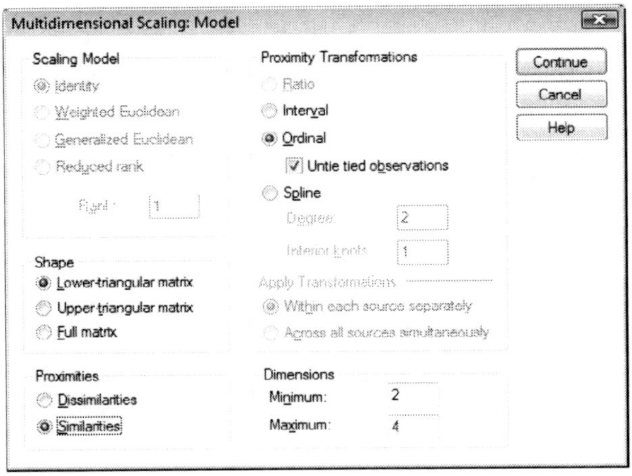

Step 6

Given that you have a single source example, any actions related to multiple sources are dimmed in the dialogue boxes. Now press '**Continue**', which returns you to the previous window. Now skip '**Restrictions**' for this simple example and choose '**Options**'.

Step 7

Now you must indicate which of the starting configuration options you wish to use in **Figure 9.7** (try several and see which is best). Choose a classical scaling start by '**Torgerson**'. For the stopping criteria, use the default settings which naturally may be changed; for example, increase the number of iterations if needed, to help convergence to a minimum stress value.

The option '**Use relaxed updates**' can be ticked, if you wish. It is a device for greater computational efficiency by the algorithm.

Press '**Continue**' and click on the '**Plots**' button.

Figure 9.7: 'Multidimensional Scaling: Options' Dialogue Box

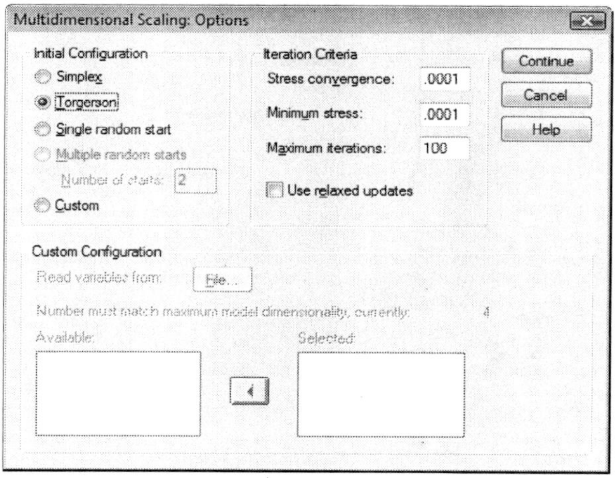

Step 8

In the plot options as shown in **Figure 9.8**, request the '**Stress**' plot, '**Common space**' (2-D plot) and the two pair-wise diagnostic plots as indicated. Press '**Continue**' and proceed to the '**Output**' window.

Figure 9.8: 'Multidimensional Scaling: Plots' Dialogue Box

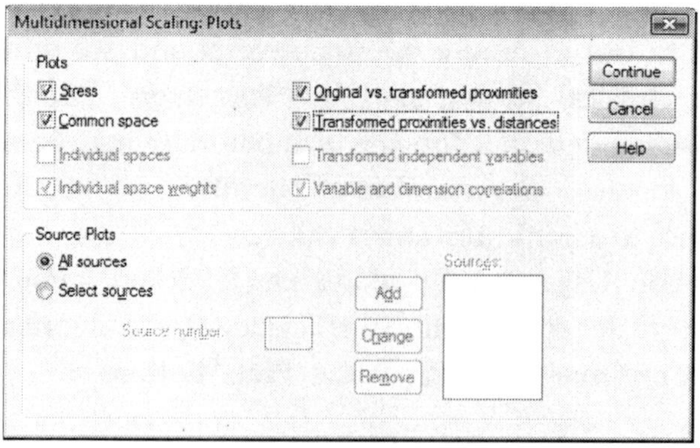

Step 9

In terms of the output choices in **Figure 9.9**, you have ticked all possible (highlighted) options so you will get the coordinates in the reduced space, the derived distances, the transformed proximities and the original inputted data. A report on the stress history and final value for the current configuration is output also, as well as the stress decomposition across objects and sources (if several). You also have the option of saving the output to file. You did not impose any restrictions on the common space. Finally, click '**Continue**' and press '**OK**' on the first window.

Figure 9.9: 'Multidimensional Scaling: Output' Dialogue Box

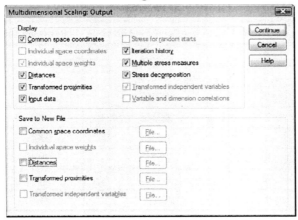

When this is done, the following output is generated, starting with a printout of the input data (**Proximities**) and the iteration history of the algorithm.

Table 9.1: The Proximities Matrix

Brands	1	2	3	4	5	6	7	8
1								
2	6.000							
3	8.000	9.000						
4	6.000	8.000	6.000					
5	2.000	3.000	1.000	1.000				
6	3.000	3.000	1.000	2.000	1.000			
7	1.000	2.000	2.000	3.000	8.000	5.000		
8	2.000	2.000	2.000	2.000	7.000	5.000	9.000	

The original data of proximities is now shown in **Table 9.1**, which confirms the original perception that brand 3 and 2 and brands 7 and 8 are the most similar pairs (largest proximity values), so hopefully this will be shown in the final 2-D plot. The iteration history shows how the stress value is reduced (for a 2-D configuration) from the initial configuration to the final configuration after 13 iterations, stopping when the stress is very small as specified in the options (**Figure 9.7** above).

Table 9.2: Iteration History

Iteration	Normalised Raw Stress	Improvement
0	.15002[a]	
1	.00656	.14346
2	.00467	.00189
3	.00366	.00100
4	.00302	.00064
5	.00257	.00045
6	.00227	.00031
7	.00205	.00022
8	.00188	.00017
9	.00174	.00014
10	.00162	.00012
11	.00151	.00011
12	.00140	.00010
13	.00131	.00009[b]
a. Stress of initial configuration: Torgerson start.		
b. The iteration process has stopped because Improvement has become less than the convergence criterion.		

Different starting configurations may lead to different final plots and stress values. Here, you have chosen a '**Torgerson start**'. You can revisit the analysis with one of the other starting configurations such as '**Simplex**' or random starts or user-supplied starting configurations, and note the different initial stress values and number of iterations needed to reach a final configuration. Hopefully, they will all yield a similar configuration.

Figure 9.10: Scree Plot of Stress

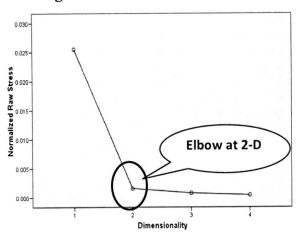

Looking at the minimum stress attainable (from a '**Torgerson start**') for each dimension from 1 to 4 (out of 8 - 1 = 7 possible), the elbow at the second dimension indicates no appreciable reduction in stress after the second dimension and, therefore, you should be happy to use a 2-D representation and proceed to look at the rest of the 2-D output, which consists of the stress report and other output as follows.

Table 9.3: Stress and Fit Measures

Normalised Raw Stress	.00131
Stress-I	.03619[a]
Stress-II	.07840[a]
S-Stress	.00286[b]
Dispersion Accounted For (DAF)	.99869
Tucker's Coefficient of Congruence	.99934
PROXSCAL minimises Normalised Raw Stress	
a. Optimal scaling factor = 1.001. b. Optimal scaling factor = 1.000.	

The final normalised raw stress value is printed as a measure of fit, as well as several other stress measures for the final 2-D configuration. You will wish to find all these close to zero, indicating a good overall fit.

At a glance, the key quantity of normalised raw stress is quite small (0.00131); indicating a good fit in a global sense. The other measures of stress available also indicate a good fit. The DAF value here is just (1.0 – raw stress) and Tucker's value is $\sqrt{(DAF)}$, so values close to 1.0 indicate a good fit.

Table 9.4: Decomposition of Normalized Raw Stress

		Source SRC_1	Mean
Object	Brand 1	.0028	.0028
	Brand 2	.0005	.0005
	Brand 3	.0015	.0015
	Brand 4	.0028	.0028
	Brand 5	.0020	.0020
	Brand 6	.0006	.0006
	Brand 7	.0002	.0002
	Brand 8	.0002	.0002
Mean		.0013	.0013

Table 9.5: Brand Map Positions

	Dimension 1	2
Brand 1	.599	.121
Brand 2	.449	-.185
Brand 3	.615	-.252
Brand 4	.488	-.055
Brand 5	-.674	-.366
Brand 6	-.146	.839
Brand 7	-.666	-.056
Brand 8	-.664	-.047

Table 9.6 details the position in Euclidean space of each point. For instance, if you were to find the Euclidean distance between brand 1 and 2, whose coordinates are given in the coordinate table as (0.599,0.121) and (0.449, -0.185), you would find $d_{12} = 0.341$, which is recorded in the distance matrix below along with all the other inter-point distances. The stress functions generally are invariant to translation and uniform dilation, so this allows some freedom in terms of normalising the configuration so that it is centred on the origin. This means (using SPSS) that the coordinates for each dimension, when summed, will total zero:

$$\sum_{i=1}^{n} x_{ir} = 0, \text{ for } r = 1, 2.$$

The decomposition of stress across each object also is detailed for each source (only one here, so the mean is the same) and so you can see where the major contributions lie.

The first of the following two tables also generated in the output is the distances between the pairs of points on the plot and hence these are the derived distances, which may not exhibit the same rank order as the original input matrix of perceived similarities.

Table 9.6: Euclidean Distances between Brands on the Map

Brands	1	2	3	4	5	6	7	8
1	.000							
2	.341	.000						
3	.373	.178	.000					
4	.208	.136	.234	.000				
5	1.363	1.138	1.294	1.203	.000			
6	1.035	1.185	1.330	1.096	1.315	.000		
7	1.277	1.123	1.296	1.154	.310	1.035	.000	
8	1.274	1.122	1.295	1.152	.319	1.026	.009	.000

Table 9.7 is the table of transformed distances. These transformed proximities are scaled by PROXSCAL to sum to $^{n}C_2$ when squared (for one source), hence their sum equals $^{8}C_2 = 28$.

Table 9.7: Transformed Distances

Brands	1	2	3	4	5	6	7	8
1	.000							
2	.341	.000						
3	.289	.158	.000					
4	.289	.158	.289	.000				
5	1.289	1.138	1.296	1.289	.000			
6	1.036	1.138	1.331	1.138	1.317	.000		
7	1.289	1.138	1.289	1.138	.289	1.036	.000	
8	1.276	1.138	1.289	1.154	.289	1.027	.009	.000

The derived distances may not be in the correct rank order, so the transformed distances are found that are as close as possible to the derived distances.

Figure 9.11: Residuals Plot

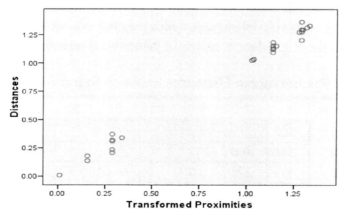

Consider the plot of map distances *versus* transformed distances in **Figure 9.11** – the inter-object distances in the chosen dimension plotted against the fitted values from the algorithm. A good fit would result in a 450 line or close to it without too much scattering. The data here is simply the values plotted in increasing order from **Tables 9.7** and **9.8**. In this case, this desired is scatter present – this is sometimes called a Shepard diagram.

Figure 9.12: Transformed Proximities Plotted against Input Proximities

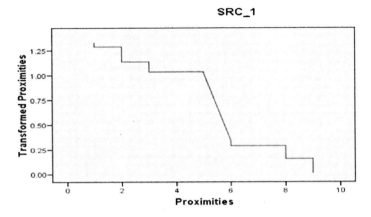

Transformation: matrix conditional ordinal (ties allowed to be untied).

Consider now the plot of transformed proximities *versus* inputted proximities – if a lot of smoothing is required, this indicates a bad fit – a lot of horizontal portions on the plot where smoothing to a monotone curve was needed. For proximity data, this is a monotone decreasing path. You allowed ties in the original input data to be untied (this is called the primary approach to ties) in the transformed proximities. Now investigate the change (if any) in the resulting stress and configuration if this option in the dialogue box is not ticked.

Next, compare the 1-D (simply repeat the process, but changing the dimension choice to '1' in the '**Model**' dialog box as shown in **Figure 9.13**) and 2-D configurations and stress levels corresponding to your choice of analysis settings.

Figure 9.13: 'Multidimensional Scaling: Model' Dialogue Box (2)

Figure 9.14: 1-D Plot of the Brands

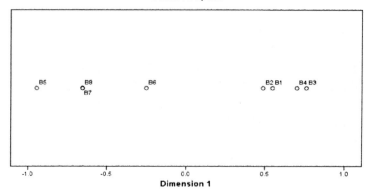

Table 9.8: 1-D Stress and Fit Measures

Dimensionality: 1

Normalised Raw Stress	.00252
Stress-I	.15975[a]
Stress-II	.28897[a]
S-Stress	.06465[b]
Dispersion Accounted For (DAF)	.97448
Tucker's Coefficient of Congruence	.98716
PROXSCAL minimises Normalised Raw Stress	
a. Optimal scaling factor = 1.026. b. Optimal scaling factor = .982.	

The 1-D output: The minimum stress value, as expected, is larger than that of the 2-D solution. As regards the plot, an essentially bipolar structure is in evidence with the brands distributed along the axes with differentiation in perception being related to relative positions. The researcher and respondent can surmise as to the underlying possible differences and perhaps the dimension may emerge as being essentially one of, say, quality or price, etc. Of more substantial interest, of course, is the 2-D plot, given that it represents in this example the optimal trade-off of fit *versus* simplicity of interpretation and insight.

The 2-D plot: The plot is your key output. Possibly, you may confirm the map structure using different MDS approaches – for example, using a metric or interval scaled transformation instead of your current ordinal approach (if the input data permits) and indeed different starting configurations. Remember also that this plot refers to an individual source (respondent) and several sources may be analysed at the individual level with multiple plots produced or the aggregate level with a common plot produced and subsequent researcher interpretation required.

Examining this plot, note how similar various brands appear to be by virtue of being closely positioned on the map. Groups or clusters of objects (brands) can be easily identified informally. For a marketer, it may be of interest to adduce the structure at an individual, or more usually at an aggregate, level. Further attempts to interpret the various dimensions may form part of this perception (perhaps subconsciously) and the relative positions of the objects may serve as a clue in this regard.

In this case, dimension 1 contrasts brands 1, 2, 4 and 3 against particularly brands 5, 7 and 8. Dimension 2 contrasts brand 6 against the rest. Knowledge of the brand features can help elicit likely dimensional interpretation. For example, if the brands are radio stations, then brands 1, 2, 3 and 4 along the first dimension might be associated with popular music/programme content aimed at the under-25 age group whilst brands 5, 7 and 8 might be middle-of-the-road music aimed at older listeners, with one station (Brand 6) offering classical and world music content catering perhaps for a niche segment. Consequently, the first dimension might be interpreted as younger/popular content *versus* older conventional content; the second (vertical) axis might be viewed as mainstream *versus* niche, etc.

Figure 9.15: Plot of Brands

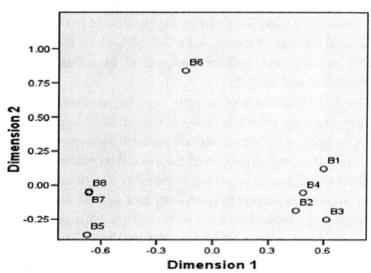

Therefore, you can visualise the similarity structure of the brands, note clusters of similar brands and assess likely dimensions of perception associated with the brands. Respondents can examine the map and offer reasons as to why the brands are differentiated in the way shown. Note also that the relative configuration is unchanged under rotation or reflection so that, as an aid to intuiting the possible meaning of the various dimensions retained in the solution, you may use a (subjectively) rotated version of the axes if it simplifies the structure.

As a simple comparison, compare the plot and fit obtained by using the option of an interval proximity transformation (if the input data were distance data, you also could use a ratio scaled metric transformation) and note if the structure is replicated.

Figure 9.16: Comparison of Interval and Ordinal MDS Plots

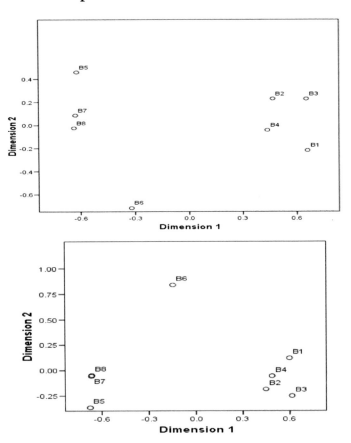

Figure 9.17: 'Multidimensional Scaling: Model' Dialogue Box (3)

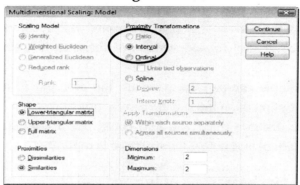

Apart from a reflection and a smaller stress (0.0013 versus 0.017) achieved using ordinal MDS, the structure is the same for both.

CONCLUSION

The MDS process with its modest data requirement may serve as an initial foray into the issue of similarity perception. The starting point is a 'distance' matrix, which can be constructed from objective attribute data or arises from a distance or proximity matrix (similarities) that are often subjectively provided by a single source (individual) or several sources.

Visualisation using MDS is useful for presenting the object set in a low dimensional space in an optimal manner using measures of fit such as 'stress' or a 'loss' measure. The researcher thus can endeavour to understand the nature of the perceived similarity, the underlying dimensions in this assessment and possible clustering patterns. Rotation of axes may facilitate the interpretation as well as use of object attributes.

This chapter focused on non-metric MDS, with input data being of an ordinal nature but, of course, metric options are readily available also and may be easily applied if the input data is ratio or interval scaled.

The analysis can be at the individual level or aggregated across a sample, assuming that the respondents are using the same two underlying dimensions with the same weightings attached. Extensions to individual scaling models such as INDSCAL (Individual Differences Euclidean Distance Model) are available in SPSS and other programs when several data matrices are being analysed.

The interested reader can access the SPSS algorithm library, case studies and syntax reference files for other options for the analysis.

REFERENCES

Cox, T.F. & Cox, M.A.A. (2001). *Multidimensional Scaling*, 2nd ed. Boca Raton, FL: Chapman and Hall/CRC.

Kruskal, J.B. (1964). 'Multidimensional Scaling by Optimizing Goodness of Fit to A Nonmetric Hypothesis', *Psychometrika* 29(1): 1-28.

Meulman, J.J. & Heiser, W.J. (2005). *SPSS Categories 14.0*. Chicago, IL: SPSS Inc.

Shepard, R.N. (1962a). 'The Analysis of Proximities: Multidimensional Scaling with An Unknown Distance Function I', *Psychometrika* 27(1): 125-140.

Shepard, R.N. (1962b). 'The Analysis of Proximities: Multidimensional Scaling with An Unknown Distance Function II', *Psychometrika* 27(2): 219-246.

Torgerson, W.S. (1958). *Theory and Method of Scaling*. New York: Wiley.

INDEX

**APPROACHES TO
QUALITATIVE RESEARCH**
THEORY & ITS PRACTICAL APPLICATION
A Guide for Dissertation Students
Edited by John Hogan, Paddy Dolan & Paul Donnelly

APPROACHES TO QUALITATIVE RESEARCH:
THEORY & ITS PRACTICAL APPLICATION:
A GUIDE FOR DISSERTATION STUDENTS

John Hogan, Paddy Dolan & Paul Donnelly (editors)

Approaches to Qualitative Research is primarily designed to be a qualitative research guidebook for undergraduate and postgraduate students undertaking dissertations as part of their course of study. As qualitative methodologies can be applied across a broad spectrum of disciplines, the book can be used by students working in any area of research from business studies to the social sciences. Although the book is highly theoretical, and methodologically rigorous, it is full of practical examples. Students should use this book as a handy reference guide and should see the chapters as mini-templates of what they should be aiming to produce themselves.

Available from Oak Tree Press in paperback and ebook formats.

OAK TREE PRESS

Oak Tree Press develops and delivers information, advice and resources for entrepreneurs and managers. It is Ireland's leading business book publisher, with an unrivalled reputation for quality titles across business, management, HR, law, marketing and enterprise topics. NuBooks is its recently-launched imprint, publishing short, focused ebooks for busy entrepreneurs and managers.

In addition, through its founder and managing director, Brian O'Kane, Oak Tree Press occupies a unique position in start-up and small business support in Ireland through its standard-setting titles, as well as training courses, mentoring and advisory services.

Oak Tree Press is comfortable across a range of communication media – print, web and training, focusing always on the effective communication of business information.

Oak Tree Press, 19 Rutland Street, Cork, Ireland.

T: + 353 21 4313855 F: + 353 21 4313496.

E: info@oaktreepress.com W: www.oaktreepress.com.

CPSIA information can be obtained at www.ICGtesting.com
Printed in the USA
LVOW10s2259140314

377476LV00005B/162/P